Living Deep Ecology

Environment and Society

Series Editor: Douglas Vakoch

As scholars examine the environmental challenges facing humanity, they increasingly recognize that solutions require a focus on the human causes and consequences of these threats, and not merely a focus on the scientific and technical issues. To meet this need, the Environment and Society series explores a broad range of topics in environmental studies from the perspectives of the social sciences and humanities. Books in this series help the reader understand contemporary environmental concerns, while offering concrete steps to address these problems.

Books in this series include both monographs and edited volumes that are grounded in the realities of ecological issues identified by the natural sciences. Our authors and contributors come from disciplines including but not limited to anthropology, architecture, area studies, communication studies, economics, ethics, gender studies, geography, history, law, pedagogy, philosophy, political science, psychology, religious studies, sociology, and theology. To foster a constructive dialogue between these researchers and environmental scientists, the Environment and Society series publishes work that is relevant to those engaged in environmental studies, while also being of interest to scholars from the author's primary discipline.

Titles in the series

Living Deep Ecology: A Bioregional Journey by Bill Devall and edited with an introduction by Sing C. Chew

Environmental and Animal Abuse Denial: Averting Our Gaze edited by Tomaž Grušovnik, Reingard Spannring,, and Karen Lykke Syse

Conservation, Sustainability, and Environmental Justice in India edited by Alok Gupta

Secular Discourse on Sin in the Anthropocene: What's Wrong With the World? by Ernst M. Conradie

Environment, Social Justice, and the Media in the Age of Anthropocene, edited by Elizabeth G. Dobbins, Maria Lucia Piga, and Luigi Manca

Global Perspectives on Eco-Aesthetics and Eco-Ethics: A Green Critique, edited by Krishanu Maiti & Soumyadeep Chakraborty

Australian Wetland Cultures: Swamp Thinking, edited by John Charles Ryan & Li Chen

Motor Vehicles, the Environment, and the Human Condition: Driving to Extinction, by Hans A. Baer

Living Deep Ecology

A Bioregional Journey

Bill Devall

Edited with an introduction by
Sing C. Chew

LEXINGTON BOOKS
Lanham • Boulder • New York • London

Published by Lexington Books
An imprint of The Rowman & Littlefield Publishing Group, Inc.
4501 Forbes Boulevard, Suite 200, Lanham, Maryland 20706
www.rowman.com

6 Tinworth Street, London SE11 5AL, United Kingdom

British Library Cataloguing in Publication Information Available

Library of Congress Cataloging-in-Publication Data
Names: Devall, Bill, 1938– author. | Chew, Sing C., writer of introduction.
Title: Living deep ecology : a bioregional journey / Bill Devall ; edited with an introduction by Sing C. Chew.
Description: Lanham : Lexington Books, [2021] | Series: Environment and society | Includes bibliographical references and index.
Identifiers: LCCN 2020041838 (print) | LCCN 2020041839 (ebook) | ISBN 9781793631862 (cloth) | ISBN 9781793631879 (epub)
Subjects: LCSH: Biogeography—California, Northern. | Biogeography—Oregon. | Biodiversity conservation. | Klamath Mountains (Calif. and Or.) | Siskiyou Mountains (Calif. and Or.)
Classification: LCC QH105.C2 D48 2021 (print) | LCC QH105.C2 (ebook) | DDC 577.2/2—dc23
LC record available at https://lccn.loc.gov/2020041838
LC ebook record available at https://lccn.loc.gov/2020041839

♾™ The paper used in this publication meets the minimum requirements of American National Standard for Information Sciences—Permanence of Paper for Printed Library Materials, ANSI/NISO Z39.48-1992.

For our three wonderful and faithful friends: Beau, Bert, and Billy

Contents

Introduction

Sing C. Chew

Over the course of his life, Bill Devall embraced the tenets of deep ecology in his writings and in the conduct of his personal life. His publications were inspired by the works of Arne Naess, Gary Snyder, and others, and in his first book, *Deep Ecology*, with George Sessions, they introduced the philosophy and practice of deep ecology to the North American audience in 1985. This first book has been widely cited as the key reference text for deep ecology, and to this day, it still remains in print. Following its publication, according to the *San Francisco Chronicle*, "Deep Ecology is subversive, but it's the kind of subversion we can use." The intention then was to reorient environmental thinking and action from one that was just reformist to a new ecological philosophy and practice. The writings of Norwegian philosopher Arne Naess (1973), especially his article entitled "The Shallow and the Deep, Long-Range Ecology Movements: A Summary," provided the philosophical insights and social practice and, according to Bill, was the orientation that was needed for the grassroots reform environmental movement to consider in North America then.

For Bill Devall, the continuing environmental crisis then was one that has to be understood as a crisis of character and of culture. What was needed was a shift from a view that is anthropocentric to one that is ecocentric. Instead of using the term biocentric, along with Warwick Fox (1990), Bill preferred the term ecocentric, as it entails being Earth-centered (the Earth as a whole), and therefore this term covers the ecosystems, the rivers, and mountains rather than just focusing on living things as the term biocentric would imply. To adopt such an ecocentric view of the world, following Naess, Bill promoted the embrace of a way of seeing the world that is ecosophical. This shift to ecocentricity does not need something new according to him, all that is required is to reawaken something very old, something he would term Earth wisdom—the dance of unity of plants, animals, humans, and the Earth. Such a development of Earth wisdom necessitates the synthesis of theory and social practice in our everyday activity. This Earth wisdom or ecosophy forms the basis for the deep ecologist's philosophical orientation. Furthermore, he insisted that this ecosophy is not unique, it can also be found repeated in other philosophies and religions all over world history, including Native American, Zen Buddhism, Taoism, the pre-Socratic Greeks, and in other rituals and cults of the many gathering and hunting tribes.

Such Earth wisdom would also require a shift in the normativity of our valuation process. Following the "platform of deep ecology" (see the preface to this book), according to Bill Devall, the well-being and flourishing of human and nonhuman life on Earth have value in themselves. In other words, there is inherent value in nonhuman life that is independent of the usefulness of the nonhuman world for human purposes. In this sense, the richness, complexity, and diversity of life-forms all contribute to these values. All species have a function in our ecosystem and are valued in themselves. No species (such as human beings) have a higher inherent worth above other nonhuman species. Therefore, as Bill Devall sees it, the complexity and diversity of the ecosystem is being destroyed by human practices undertaken to meet the ends and needs of human communities all at the expense of the health of the ecosystem. Such degradative actions on the part of humans call for direct actions from grassroots environmental movements so that the ecosystem can be safeguarded.

Protecting the health and sustainability of the ecosystem requires our consideration of human consumption and the issue of unceasing exponential population growth, which have a tremendous impact on the future of the planet. Bill Devall noted this population issue then, and shared Arne Naess's suggestion that human communities can flourish and have a rich life, not necessary an extravagant life, by having a substantially smaller population. To Bill, the richness of culture such as literature, art, and compassionate human relations do not require a large number of human beings. The goal is to have ecologically sustainable communities so that everything including nonliving things can flourish.

Bill Devall's subsequent books, *Simple in Means and Rich in Ends* (1988) and *Living Richly in an Age of Limits* (1993), translated these ideas into an evaluation and discussion of the existing environmental crisis and identified practical solutions and direct actions taken by mostly grassroots environmental movements. *Simple in Means and Rich in Ends*, whose subtitle was *Practicing Deep Ecology*, elaborated on the themes of his first book by outlining strategies on how the philosophical and practical actions of deep ecology can be considered and undertaken. It covered a discussion of the ecological self, gender, literature, poetry, and music from a deep ecology position. From the level of the individual to that of the community, a discussion of the land community following the writings of Aldo Leopold (1949) was undertaken. In *Simple in Means and Rich in Ends*, we begin to see the introduction of issues and ecological concepts from mostly North American environmental writers to enrich the analysis that deep ecology with its Norwegian origin can offer. *Simple in Means and Rich in Ends* and *Living Richly in an Age of Limits* covered the broad environmental themes and those of the "platform of deep ecology." When we compare these two books to this last book, *Living Deep Ecology*, they include fewer personal recollections of his conservation experiences. Instead, the two books read more like reporting commentaries on the environmental issues of the time spliced with some personal reflections and experiences.

Living Richly in an Age of Limits: Using Deep Ecology for an Abundant Life was written as a manifesto for North America's middle class. In this book, the focus is on the lifestyle of North Americans that to Bill was excessive and causing extreme damage to the ecosystems. The book's objective was to alert North Americans to their responsibility of decreasing their exuberant life styles thereby reducing the consumption of the natural resources of the planet, notwithstanding the destruction of the landscape and river systems and the loss of biodiversity that was already taking place. The slogan, popular in the northcoast region of California, "live simply so that others can simply live" strikes a note here. Richness of life does not have to be excessive consumption at the expense of other living and nonliving things. Buddhism's middle way was offered as a path for social practice to follow. Mindful consideration to the precepts, and attention to the consequences of our action when we search for the way to undertake appropriate actions in certain specific circumstances are what Bill Devall suggested that we should consider. Such practices, should according to him, broaden our identification with Nature and narrow the gap in our historical endeavors of reproducing human communities in the late twentieth century. To date, this gap remains true today especially in an era of global warming, biodiversity losses, natural resource scarcities, and landscape devastation.

How do we live an abundant and rich life without continuing to put our world and Nature on the unsustainable path we are on when *Living Richly in an Age of Limits* was published. By then there was a wide variety of literature by environmentalists and scholars on how this can be done. Bill's efforts in this was not to question nor downplay their suggestions and contributions but to offer a nuanced set of insights and suggestions utilizing deep ecology to fashion the paths ahead.

To regain our identification with Nature, he suggested that Nature has to be experienced. The continuing march of urbanization in North America has left the people totally distanced from Nature. A set of strategies was offered on how to live in an urban environment without losing touch of being in Nature. Social practices on how to live in cities or highly urbanized areas were sketched out. Cities are places with high population levels located on a landscape with towering buildings, power grids, highways, and streets, which have transformed our whole ecosystem.

Relying on the work of Theodore Roszak (1992), Bill Devall declared that living in such a megacity we find that our ecological unconscious being shut off as a result of our human egoistic domination. Such a condition further widens our separation from Nature in which we are a part of. Can we promote the wildness of Nature in our conscious practice of deep-ecology living in a highly urbanized community? To this, Bill Devall replied that we should try and, as well, remember what it is like. What needs to be realized is to make the highly urbanized area hospitable not only to humans but also to other native living things such as native plants, animals, and migratory birds that used to inhabit the urban area many years ago prior to urbanization. Protection of open spaces, minimal alteration of rivers and streams that traverse the landscape and wetlands, spaces for each household to plant food, and the integration of the countryside with the city were a long list of suggestions offered. The efforts should be to focus on mindfulness whereby individuals seek to remember the wildness that used to permeate the location where humans lived in the megacity. Explore degraded wild lands, rivers, and marshes to re-experience what has been lost. Try to remember the experience of what is under the concrete of the human landscape in which the individual lives in a megacity. Dream of the ravens and speak up for the coyote and bear that used to roam this now humanized landscape. Celebrate the seasons that often are harder to see and experience in a concrete landscape with the natural environment changing according to the weather and the different smells associated with each season. Discover the streams and their courses to the seas and oceans where they once drained. Demand the right to walk in a megacity, which is a subversive activity as mechanized vehicles are the norm of transportation. Finally, and not exhaustively, oppose further development of any kind to reduce the large human footprint that already has taken place in the course of the urbanization process.

Besides the need to rediscover our ecological self if we live in an urbanized environment, our ecological consciousness can be deepened with visits to areas of landscape degradation such as the massive deforestation occurring all over the world to fuel the human consumption for an exuberant lifestyle. For North Americans, the massive clear-cuts of the ancient forests of the Pacific Northwest have to be realized by visiting and experiencing the loss of the forests and all the other living beings residing in the forested landscape. We need to bear witness to the landscape devastation as a result of the massive deforestation occurring in the Pacific Northwest and in other parts of the world such as the Amazon rain forests. To Bill Devall, Nature and other living things are so forgiving in spite of the recklessness that the human community has inflicted on them for centuries. As he wrote, they continue to give forever.

The American environmentalist David Brower's dictum "let the mountains talk and the rivers run" was embraced warmly by Bill Devall so that we can return to Nature and reawaken our identification with it. For further practice, experiential education to awake us from our collective suicidal path was proposed as a way to develop this ecocentric consciousness. Pedagogically, it was taught in workshops and environmental classes in the universities and institutes such as Humboldt State University (HSU), University of California—Santa Cruz, and California Institute of Integral Studies. Workshops such as the "Council of All Beings," "Nuclear Guardianship," "Experiencing our own Biocreations," and "Mountains and Rivers Session" as experiential encounters can help us to "become aware of, acknowledge, and integrate into our awareness our anxiety about the state of the earth and our deep love for our home" (Devall 1988).

PLACE AND LIFE

The exploration of our evolving relationship with where we live and our life experiences in the place underlines the structure of this last authored book of Bill Devall.

In *Living Deep Ecology,* Bill Devall covers the various aspects of bioregional living and practices that he witnessed and lived through. In this sense, the book is unique as it combines lived-experiences or *lebenswelt* with the concept of *bioregionalism* in sketching out alternate practical ways of relating to Nature and the transformations of the physical and urban landscapes. Bioregionalism reflects a vision of life practices and the proponents of this world-view came from diverse disciplines and professions. Poet Gary Snyder wrote early treatises on the idea of bioregionalism (Snyder 1974, 1980). Peter Berg, actor and writer, of the Planet Drum Foundation was also another such proponent of the vision (Berg 1978, 2009). Others such as Jim Dodge (1981), Freeman House (1998), and Kirkpatrick Sale (1991) also added to this perspective.

What does this vision and life practice cover? Like the tenets of deep ecology, bioregionalism's basic premise is that the human being is part of a living world, and as a result there should not be a separation between the human and other living things. This interrelationship between humans and their surroundings is considered harmonious, complex, and diverse—in short, a view of the world that is Gaean. What the bioregional vision calls for is a radical change in our relationship to our natural surroundings and our life practices. It is in contrast to the twentieth century and to even today, an era where you find the following: (a) excessive consumption of natural resources, (b) impending scarcity of these resources and the overdrawing of them by humans leading to species extinction, (c) uncontrollable urbanization, (d) excessive deforestation, and (e) climate change.

Such a counter vision requires a different scale of life practice. Whereas the common or orthodox assumption of scale operates at the level of the state, and perhaps nation/global, the bioregional scale has region in place of state, and community replacing nation/world (Sale 1991). This shift in juridical-spatial dimension assumes that for bioregionalism, the emphasis is on localism instead of the globalism that was commonly assumed then and continues to be encouraged even today. If the scale shifts to localism, then the focus is on the concept of place and how this informs our life practice (Aberley 1998; Carr 2004; Dodge 1981; House 1998; Sale 1991; Berg 1978; Berg and Dasmann 1978; Chew 1997, 2008; Devall 1998; Thayer 2003). Given this, the conduct of life practice means to bear responsibility to the locale where one inhabits. Such an obligation entails that essential life and natural processes of the place remain intact and evolving. In this sense, place is then considered to be the bioregion, with bios meaning life and region meaning life-territory (Dodge 1981). In such a context, life and place are intertwined (Thayer 2003).

By pursuing this sense of place, one also develops a direct relationship with one's immediate surroundings. This relationship is then transformed into a set of life practices and awareness. It involves being aware of the rhythms and contours of the landscape and the ecological relations that underline its dynamics. It entails being aware of the diversity of life and the need either to restore what has been degraded or to ensure that ecological degradation does not take place. What it also means is the enriching of the sociocultural life and the natural aspects of the place.

To achieve this requires the efforts of not only the development of an individual understanding and realization but that of the whole community. The emphasis and responsibility then is on the community, and for the community to partake in such life practices over the duration of generations of community members. In this light, it is a call for a return to the community that has been forsaken lately for individual ego-centrality—the result of the manner in which our socioeconomic and political behaviors have evolved to date.

This realignment of socioeconomic and political orientation to community redraws and realigns in spatial dimensions the parameters of a bioregion to be defined not by hard political boundaries that we have resided under such as state, country, or nation but to soft boundaries. These soft boundaries are to be determined by physical and biotic shift criteria such as changes in plant and animal life demarcating the boundaries of one bioregion with another

(Dodge 1981). Coupled with landforms, such as watersheds and mountain ranges, these soft boundaries would add to the demarcation boundary of a bioregion. Added to this, we can also consider cultural and phenomenological aspects of social life located within these soft boundaries.

If socioeconomic and political life are determined and defined by such boundary dimensions that are closely aligned to landscape and rhythms of natural life, the bioregion vision is shaped to counter what is viewed as sources and causes of the destruction of Nature, and the strain and tension of sociocultural relations. It is proposed as the counter alternative to our existing and practicing political economy and cultural activities. Sale (1991) and Carr (2004) offer an articulation of the bioregional vision in these spheres governing our social life. In the realm of economics, the bioregional view is that the economy is based on maintaining the natural instead of unsustainable excessive extraction of natural resources—in other words, an economy based on conservation. What this means is an economy dependent on "a minimum number of goods and the minimum amount of environmental disruption along with the maximum use of renewable resources and the maximum use of human labor and ingenuity" (Sale 1991:69). In this view, maximal economic grow is not deemed to be the goal of such an economy; instead, sustainability is preferred.

With conservation as an operating principle of a bioregional economy, self-sufficiency of the economy becomes the underlying principle. A self-sufficient bioregion would not only be economically stable; it would also mean that it will not have to be so dependent on other regions or other parts of the world to ensure its economic reproducibility. By avoiding this, a bioregional economy would not be destructive to natural areas and sociocultural lifestyles of other regions and parts of the world, as the self-sufficient bioregion would not be so dependent on these resources. With a self-sufficient bioregion based on the aforementioned economic precepts, it would mean that all the issues and tragedies and consequences of a growth-based economy ranging from pollution, diseases, toxicity, and crime could be avoided.

As the focus of this view is shifted to the level of community and place, in the area of governance, decentralization as opposed to centralization (a feature of our current global world) is the political norm. Where diversity of all life and interdependency relationships are celebrated, the emphasis is not on hierarchical efficiency. Rather, complementarity and reciprocity of relationships are fostered. What this entails is that there is no identification of hierarchical priority of functions or rank, for all parts are necessary to the reproduction of the whole. Shared responsibility becomes the norm of governance. With this form, governance, communication, and networking become necessary and important. Networks therefore ground the political governance to horizontal decision-making and autonomous units of political participation. Given the celebration of diversity of life, it would seem likely that different political relations would likely exist according to their own environmental settings and their own political needs. Instead of political homogenization that is the tendency of political globalization strategies today, the stress is the autonomous evolution of political practices that are compatible to life practices and the requirements of the cultural-economic landscape.

If this becomes the nature of political and economic governance, the sociocultural aspects of life would entail certain specific characteristics that embrace these thematics. The principal of intrinsic right to evolve for all living things suggests that the diversity of culture and other life-forms is encouraged. With the linkage to place, the cultural understanding of historical relationship between the human community and the land is understood and remembered. Cultural practices and transformation thus evolve and adapt to the biological integrities of the landscape.

Particularly, in this era of global environmental, socioeconomic, and political crises, this book offers the reader possible outcomes to mediate through the problems and changes we face today. With the ongoing deurbanization and reurbanization of our physical landscapes, along with the fading of traditional lifestyles with its replacement today of a global digitalized public commons, this book reports on the social and cultural beliefs and practices that were undertaken in northern California from the 1970s onward. Although grounded mainly in the Klamath-Siskiyou bioregion, the chapters of this book examine issues and topics that are relevant to any person interested in our evolving relationship with Nature. It identifies social, economic, and ecological practices and attempts to develop sustainable and resilient communities.

At the social level, one can try to address the bioregional concept raised in this book in the areas of urban and rural partnerships that are increasingly necessary in an era of global climate changes, especially on issues concerning food production, social identities, and political governance and actions. The opportunity for concretizing social, economic, and ecological imaginaries can be undertaken leading to the development of strategic directions and priorities.

By no means was Bill Devall's effort to saving Nature only devoted to authoring books. As a deep ecologist, he was involved in the practice of conservation and environmental action at both the local and national levels. At the local level, he was a founding member of the North Coast Environmental Center based in Arcata (California) and was very active in efforts to establish recycling and the protection of the local beaches, forests, and endangered species. Nationally, he was actively involved in the protection of the ancient forests of the Pacific Northwest. He was frequently the target of anti-environmentalists in their writings. Alston Chase's book, *In a Dark Wood*, repeatedly identified Bill Devall as one of the main political ideologists of the radical environmental movement coupling him with the founder of Earth First!, Dave Foreman. Bill Devall's accounting of "redwood summer"—a summer of blocking access to the ancient forests of northern California and lumber production—in 1990 bear witness to a national campaign protesting the clear cutting of our ancient forests. What happened historically over the varied attempts to protect the ancient forests in the Pacific Northwest can be found in several chapters of this book. The national campaign led to the Sierra Club's publication of a pictorial book, *Clear Cut: The Tragedy of Industrial Forestry* (1995), edited by Bill Devall.

After these many years of bioregional living and exploration in the Klamath-Siskiyou bioregion, among the notes and papers Bill Devall left behind, the following quote summarizes the ecology of wisdom that he came to realize in his long journey:

> Going into the landscape, the bioregion, arouses our consciousness and out of consciousness comes conscience, our sense of moral responsibility to a place. Is this place an extension of my *self*, or is my *self* an extension of this place? I take responsibility for the integrity of this place because I am intimate with it, because it gives me solace, because in my solitude I am surrounded with aspects of my *self*—rocks off shore, redwood forests, egrets, bear, elk, alders, and the myriad of other creatures who dwell herein. I have experienced great sorrow and sadness witnessing the clearcuts, siltation from logging roads, and destruction of forests. I have engaged in numerous actions of protest, to protect the integrity of this landscape, but what am I really attempting to save? I can't save the present state of the landscape because impermanence is a fact of living. I can't even save my material body because I am a mortal being. I realize each day how fragile my mortal body has become. No, salvation means saving meaning in a world where nihilism has become the great god. I am saving my ability to find meaning in a world of fact and to savor joy in a world of sorrow.
>
> I can't "save" anything, not even my mortal body. However, in attempting to "save" the integrity of ancient forests I am "saving" the possibility for evolution to continue.

In what follows, we can hear Bill Devall's voice again.

Sing C. Chew

Preface

The slogan "Think globally, act locally" was popular during the 1970s. Like most slogans, this one is vague. What do we think about? How should we act? This collection of chapters explores these general questions in the context of real-world choices, real-world conflicts in our relationship with nature within the context of a specific bioregion—the Klamath-Siskiyou bioregion.

Numerous writers have addressed the interacting processes of globalization, global warming, post-oil social collapse, and the rising rate of species extinction due to human-caused destruction of species habitat. During this era of global warming, a spectrum of possibilities for the Earth and for international regions have been presented in the report of the International Panel on Climate Change (2007).

These global assessments, however, do not provide microspecific examples of how local communities of people can change over the twenty-first century based on the principle of "ecological sustainability."

The chapters in this book address questions that I think many people are asking in the context of their own evolving relationship with *place,* in the context of massive global changes during the twenty-first century. *Place* is more than landscape, more than descriptions of landforms presented by scientists. *Place* is the lifeworld of experience. By exploring the *place* wherein we dwell, we discover paradoxes and questions which draw us inward, into an affirmation of life that we are always in process of discovering.

Near the end of his life, John Muir was asked what he had accomplished in his years of ecological fieldwork in the Sierra. He replied, "I only went out for a walk, and concluded to stay until sundown, for going out, I discovered I was going in."

Muir was an ecological fieldworker, emotionally, physically, and intellectually involved in *place.* Muir embodied his evolving relationship with the

Sierra by walking in the mountains and then reflecting on public policy and political action. I feel it is important that we return to the roots of the conservation movement, to celebration, affirmation, witnessing for the integrity of rivers, forests, and mountains, before we enter the dangerous world of politics. Although there is considerable discussion of political campaigns in the following chapters, there is also discussion of the importance of grassroots activism based on our feelings, our sense of *place*.

Place is identified as bioregion in the following chapters. *Bio* means life and *region* means the territory or lifeworld of experience. Literally, then, a bioregion is the territory of lived experience, the territory with which a person, or a community of people, identifies as part of their self.

Some people identify their bioregion in terms of a watershed, or portion of a watershed, within which they dwell. For example, the Yurok identified with the Klamath River from its mouth upriver to its confluence with the Trinity River. The Hoopa identified with the Trinity River from its confluence with the South Fork, Trinity River to its confluence with the Klamath River. Other people consider their bioregion to include several watersheds draining from a range of mountains.

Bioregions have soft boundaries as opposed to the hard, political boundaries of counties, cities, states, and nations. Deep, bioregional culture refers to the roots of beliefs and code of conduct that determine how an individual, or a community, will give meaning to their relationship with the landscape and how the community will craft a way of life that is sustained within the limits suggested by *place*. Some writers refer to the process of crafting such a way of life as *reinhabitation*.

Some writers suggest that the Pacific Northwest bioregion begins in Humboldt County and extends through the temperate rainforests of the Pacific Northwest to Southeastern Alaska. Writers refer to this bioregion as Cascadia. For others, Cascadia includes all the rivers of California, Oregon, Washington, Idaho, British Columbia, and Alaska in which salmon migrate and spawn. Salmon, considered by many to be the totem animal of the bioregion, draw together the mountains and the ocean through their seasonal migrations.

In my own explorations of *place*, I have most identified with the territory between Mount Shasta, on the east, to the mouth of the Smith River on the north and the headwaters of the Eel River on the south. Some government planning agencies refer to this territory as the "northcoast basins" or the "Klamath-Siskiyou province." The "northcoast basins" refers to the rivers that arise in the interior of California and southern Oregon and flow to the ocean primarily through the Smith, Klamath, Mad, and Eel river systems. The bioregion used as an example in this book generally is within the boundaries of Mendocino, Humboldt, Del Norte, Trinity, and Klamath counties of California.

The chapters in this book envision possibilities for sustained commitment to healing in and with the landscape within which we dwell. In the context of the Klamath-Siskiyou bioregion that means healing ecological communities from the ravages of industrial logging, over-fishing, roadbuilding, and grazing which it has endured over the past two centuries, as well as adapting to global warming and post-cheap oil economy.

Throughout this book I use the term "conservation movement" rather than "environment movement" or "environmental justice movement" or "ecology movement."

Conservation implies taking actions based on our conscience. A conservationist devotes his or her life to caring for and dwelling within *place*.

Conservation activists discussed in these chapters are primarily grassroots activists defending the integrity of the *place* wherein they dwell. When people dwelling within their bioregion defend the integrity of their bioregion and resist political and economic forces that threaten the bioregion, they develop a sense of solidarity and a style of politics that is strong, sustained, powerful, and spiritually empowering for participants.

No doubt there will continue to be cultural wars between various people who espouse different theories of the root causes of our current problems and over extraction and distribution of natural resources among different groups in society. There will be cultural wars over the distribution of natural resources between present and future generations of humans, between women and men, between different ethnic and religious groups, between social classes.

However, in the emerging bioregional culture, these cultural wars will be grounded in listening to the vital needs of *place*. Thus in this book there are re-occurring themes of affirmation, witnessing for the integrity of *place*, defense of the refugia of salmon, old-growth redwoods, and endangered plants.

The greatest lesson I learned, in my explorations of this region, is the continuing, astonishing generosity of Nature. After all the abuse of the great redwood forests, abuse of mountains and rivers in the name of economic development over the past hundred and fifty years in California, the forests and rivers and mountains freely give themselves, give their gifts to all other sentient beings. Giant trees give themselves to loggers who convert them into lumber for the marketplace just as they give themselves to birds who nest in their limbs and to windstorms that smash them on the ground. And even when they are blown down in storms, the decaying trunks of giant redwood trees continue to give themselves over hundreds of years as habitat and shelter for a myriad of evolving beings.

As an ecological fieldworker I feel that I become part of that generosity. I freely gave myself because that which I was giving *to* is part of my self. Participating in the life of the forests and watersheds, I witness for the integrity of something greater than my narrow ego.

In the context of bioregional culture, by committing ourselves to a specific *place*, we allow ourselves to attempt to rediscover ancient wisdom. We may fail individually and collectively as a society, in our quest, but at least we will take the first steps on our journey in the faith that the generosity of Nature will provide us with meaning in a world where nihilism runs like a mad dog through a playground filled with innocent children.

True, there are many people attracted to other planets and to the heavens. Some people search for transcendental salvation and an end of history. In contrast, this book is premised on the belief stated so eloquently by poet/philosopher Gary Snyder that as human animals, we are at our best when we are singing and laughing around our own little watering hole in deep space. More and more people are discovering their own watering hole in the context of their specific bioregion. We are always coming home. Home is the intimate *place* wherein we dwell, and it is to that *place* to which we return again and again.

We return to the *place* wherein we dwell in the context of the deep, long-range ecology movement.

The "platform" of the deep, long-range ecology movement is articulated in a series of statements upon which people from different religions and philosophies can agree.

1. All living beings have intrinsic value.
2. The diversity and richness of life has intrinsic value.
3. Except to satisfy vital human needs, mankind does not have a right to reduce this diversity and this richness.
4. It would be better for human beings if there were fewer of them, and much better for other living creatures.
5. Today the extent and nature of human interference in the various ecosystems are not sustainable, and lack of sustainability is rising.
6. Decisive improvement requires considerable change: social, economic, technological, and ideological.
7. An ideological change would essentially entail seeking a better quality of life rather than a raised standard of living.
8. Those who accept the aforementioned points are responsible for trying to contribute directly and indirectly to the realization of the necessary changes.

This book is an extended comment on the application of these principles in the context of one bioregion. The philosophical literature on the deep, long-range ecology movement is extensive. This book provides the reader with suggestions on practical applications and lifestyles within one specific bioregion.

Bill Devall
Parker Creek, Humboldt County, Klamath-Siskiyou Bioregion

Chapter 1

Biogeography of the Klamath-Siskiyou Bioregion

A convergence of energy continues to shape the form and flow of Klamath-Siskiyou region.[1] Storms, generated in the north Pacific Ocean, bring winter rain along the coast and snow in the mountains. Although the highest peak in the region—Thompson Peak in the Trinity Alps—is barely 9,000 feet elevation and the highest peak along the coastline—King Peak—is barely 4,000 feet in elevation, the landscape is cut with many ridges and mountain peaks that are 4,000 feet to 6,000 feet in elevation. Snow can accumulate in the mountains for six months leading to extensive runoff during the Spring and early Summer.

Offshore of Cape Mendocino, the most westerly cape in the lower forty-eight states, are complex interrelationships of fault lines. The Cascadia subduction zone, located along the Pacific Northwest coast from Cape Mendocino to Vancouver Island, British Columbia, has only been identified within the last several decades.

The Cape Mendocino region is one of the most seismically active areas in the United States. Between 1980 and 1998 there were five earthquakes close to magnitude 7 on the Richter scale and an additional six of magnitude 6 or larger. A particularly strong earthquake occurred near Cape Mendocino on April 25, 1992, with a magnitude of 7.1.[2]

The triple juncture, approximately thirty miles offshore from the mouth of the Mattole River, is a subduction zone where one plate moves underneath another creating the huge pressures necessary to create the earthquakes in the northcoast basins. Some geologists suggest that a large earthquake in about 1700 created the shape of Humboldt Bay.

In this geologically active landscape with mild, rainy winters along the coast and deep snows in the mountains, and with a distinctive dry season from about May until October, a wide variety of plant and animal communities evolved over millions of years. Indeed, botanists during the 1930s declared that the Kalmiopsis region of the Klamath-Siskiyou mountains had one of the highest diversity of plants of any region in the United States.

The coastal redwood forests, for which the northcoast basins are particularly famous, occupy only about a thirty mile wide band along the lowlands of the northcoast rivers. Outside of the coastal redwood forests, complex forests—some dominated by oak, some by Douglas fir, madrone, Port Orford cedar, white fir, and tan oak—are found in various areas within the region.

Offshore, California gray whales migrate along this coastline from lagoons in Baja California, where they spend the winter, to feed in the Gulf of Alaska each Spring and back again in the Fall. Salmon and steelhead (an ocean-going trout) migrate back and forth from their spawning areas in streams high in the mountains into the Pacific Ocean and back again.

The seasonal movement of salmon has been a defining event for Indian tribes dwelling along the Klamath, Trinity, and other rivers of the region for thousands of years. To some extent all of the tribes inhabiting the northcoast basins—Yurok, Chilula, Hupa, Karok, Tolowa, and Wiyot—fed on salmon.

When the Spanish arrived in June 1775, at a headland they named Trinidad, they discovered a rich civilization of tribes dwelling in these river basins and along the coastline. What is now called California contained over 500 independent republics when the Spanish began their invasions during the early 1770s. Chilulas dwelled in Redwood Creek, Hoopa, along the Trinity River, Yuroks, on the lower Klamath River and Karoks on the upper Klamath, Sinkyone, dwelled along the Mattole River and the coastal zones of the King Range. Tsurai (Trinidad) prospered for at least eight hundred years as a Yurok village.

The rich cultural diversity of the region is matched by the rich biodiversity of the region. When the Spanish first viewed these coastal areas, they were rich with old-growth redwoods, bear, mountain lion, salmon, steelhead, and a myriad of co-evolving native species.

Neither the Spanish nor Russians planted settlements in the northcoast basins region, and Europeans did not begin arriving until the gold rush of 1849. Thus, the major human-caused changes by industrial civilization have occurred within the last hundred and fifty years.

The northcoast basins region, like other bioregions, is a natural system with semipermeable boundaries. The patterns of relationships of energy input and output through these boundaries defines the system. In the process of maintaining its integrity, a natural system is constantly in a state of flux and flow. Adjustments in relationships are constantly occurring. Assessments of the state of the system provide a benchmark by which to measure trends that are occurring within the system. Scientists tell us that we can never have complete information on the state of natural systems. Indeed, one of the laws of ecology is that nature is more complex than we know and perhaps more complex than we can know. However, we can use the best available knowledge, including results of scientific research, our own observations, and our intuition, to develop maps of the state of the system.

By all widely accepted scientific indicators—including the number of endemic species that are considered threatened or endangered by possible extinction due to human actions, decline of fisheries, and erosion from human-created roads and logging operations—the Klamath-Siskiyou region is in a stressed state at the beginning of the twenty-first century.

In ecological terms, the bioregion has experienced one hundred and fifty years of drawdown. Drawdown means the removal of the richness of biomass and decrease in native biodiversity that existed in the region prior to 1850.

Virtually all watersheds outside of federally designated Wilderness areas have been stripped of their mature forests. In 1850, there were an estimated two million acres of ancient redwood forests. In 2007, only approximately 5 percent of the ancient redwoods remain standing. The rest have been logged.

Another example of drawdown was presented by The Environmental Protection and Information Center (EPIC), with offices in Garberville. EPIC compiled a timeline of the decline of coho in California and the reluctance of federal and state agencies to fulfill their mandate to protect the "public trust" interest in salmon.

In the 1940s, naturally spawning coho salmon in California numbered between 200,000 and 500,000.

By the early 1990s, naturally spawning coho had fallen to between 10,000 and 15,000. Coho had been eliminated from 50 percent of streams where they were historically present.

During the early 1990s, the National Marine Fisheries Service (NMFS) rejected numerous petitions to list coho salmon in California streams under provisions of the federal Endangered Species Act (ESA).

Finally, in July 1995, under court order, the NMFS proposed to list California coho as "threatened" under the ESA.

In October 1996, under court order, the NMFS listed California's southern coho populations.

In December 1996, the NMFS withdrew its commitment to publish guidelines for conservation of coho. The agency also refused to review Timber Harvest Plans for coho impacts.

In April 1997, the NMFS, under court order, listed California's northern Csoho populations as "threatened" under provisions of the ESA.

In January 1998, EPIC filed a lawsuit in federal court alleging that federal agencies were not fulfilling their responsibilities under the ESA to protect coho salmon.

During Spring 2007, representatives of the Yurok nation and allies in the conservation movement traveled to Omaha, Nebraska, for the annual meeting of Berkshire Hatheway. They traveled to the annual meeting of this corporate giant because it owns two hydroelectric generating dams located on an upper tributary of the Klamath River in the state of Oregon. These dams block salmon from historic spawning grounds. The dams generate a small amount of electricity. Removal of these privately owned dams would help the long-term restoration of salmon runs in the Klamath River basin. The conflict between generating electricity and salmon as historic food source for the Indian nations illustrates the continuing conflict between energy and power, local and national institutions in the Klamath-Siskiyou region. A federal agency controls the licensing of continued operation of the dams on the upper Klamath River.

A major paradox of the work of local conservation activists over the past three decades, has been the tension between appeals for grassroots change, for a bioregional politics and way of life, and appeals for greater federal intervention in the region. Federal regulatory agencies and federal land management agencies such as the U.S. Forest Service are major players in all land use decisions concerning the region.

In many respects the Klamath-Siskiyou region, over the past 150 years, has been a colony of national, and now increasingly globalized, economy. Between 1850 and 1870 virtually all the lowland forests in the region were privatized, removed from public domain into private property. Nineteenth-century timber barons built an infrastructure of private railroads, shipping lines, and lumber mills to extract larger and larger amounts of timber for sale in urban markets. The logging corporations imported labor and capital and built their own company towns including Scotia and Crannell in Humboldt County. From the 1850s until 1888, Chinese laborers were employed in Eureka and in the gold mines in Trinity County. The Chinese Exclusion Act of 1888 led to the removal of Chinese to San Francisco.

Under the California Indian Act of 1872, the Hoopa and Yurok reservation was created and all other Indians residing along the northcoast of California were removed to a reservation at Round Valley in Mendocino County.

The loggers were primarily Caucasians, some who arrived from logging communities in the upper Midwest of the United States.

In the late nineteenth century it took a crew of men a week to cut one ancient redwood tree. Some people, at that time, said we would never run out of old-growth timber. But the invention of the chainsaw and diesel tractors and trucks made it possible for faster and faster extraction of more and more trees to supply an ever-increasing demand of a growing urban population.

The post-World War II housing boom increased demand for timber and with chainsaws, bulldozers, and logging trucks, loggers were able to clear-cut most of the old-growth outside of federal lands in the region.[3]

By 1965, when two large logging corporations opened pulp mills on Humboldt Bay, the handwriting was on the wall. Corporations planned to log all old-growth forests in the region by the year 2000. Raymond Dasmann, a former professor at HSU, located in Arcata, California, published his assessment of the state of the natural and cultural systems of California in 1965, *The Destruction of California*. He concluded that California was overpopulated. California then had ten million residents. In 1998, the population of California was thirty-two million and some forecasters project a population of forty-one million by the year 2020. Hispanics continue to illegally cross the border from Mexico into the United States. The number of Hispanics in the Klamath-Siskiyou bioregion is increasing.

Since 1969, when the Congress passed the National Environmental Policy Act (NEPA), federal agencies were given more power in developing management plans for river systems and forests. Although some of the federal legislation passed in the early 1970s, including the federal ESA, are quite powerful, many federal agencies have been reluctant to utilize their powers under these acts. Grassroots conservation activists were a primary social force getting government agencies to do their job of enforcing state and federal environmental laws for the general welfare. Using lawsuits, demonstrations, nonviolent direct action, public relations, and all the other tools in their toolbox, local conservation groups, frequently in alliance with national conservation groups and against great odds, have encouraged social changes in the region.

I hope we can gain insights into the social forces that will be needed over the next century to move from an economy of drawdown and greed to one where human communities share their place with other inhabitants, fully recognizing that sharing involves, even requires, a sense of limits to human enterprise. Rich cultures, as philosopher Arne Naess says, are not necessarily big cultures. Greatness and bigness are not the same. Rich cultures can be rich in a lifeworld of experience and can appreciate the greatness of wild nature. A culture rich in ceremony and in traditions that keep the world in balance recognizes intuitively the meaning of Henry David Thoreau's conclusion that "in wildness is the preservation of the world."

NOTES

1. A similar study of a community utilizing the bioregional concept in another part of California can be found in Thayer (2003). Bioregionalism as a movement started in the 1960s and 1970s in North America. There are other similar types of movements in other parts of the world, such as in India (e.g., Bahaguna 1996). In America, bioregionalism's rise was in reaction to the socioeconomic, political, and ecological

conditions of the times. The excesses of federal-state control, the war in Vietnam, the corporate control of economic life, and the technologization of life in general were some of the factors that led to the rise of this movement. This was followed with the back-to-the-land initiatives to organize life that is in contrast to the orthodox practices then, and to return to Nature (Aberley 1998; Hansson 2003). The early proponents of bioregionalism came from different disciplines and professions. Poet Gary Snyder (1974), 1980) wrote on the idea. Peter Berg (1978, 2009), actor and writer, of the Planet Drum Foundation was an early proponent. Raymond Dasmann with Peter Berg (1978) wrote on reinhabiting California that followed the bioregional themes. Other early supporters include Jim Dodge (1981), Freeman House (1998), and Kirkpatrick Sale (1991).

2. Geologist Lori Dengler (1992, 1995) has done various studies of the seismic events that have occurred in this part of Northern California. Earthquakes occur quite frequently offshore, and depending on the location of the seismic activity, to dwellers in this region, it is "a fact of life" if the magnitude is not above 4.0 on the Richter scale.

3. For a pictorial presentation of clear-cutting ancient forests in North America, see *Clearcut: The Tragedy of Industrial Forestry* edited by Bill Devall (1995). Alternative responses to clear-cutting of the forests is the practice of ecoforestry that is geared to forest sustainability, see Drengson and Taylor (1997) and Maser (1994).

Chapter 2

Passing through the Redwood Curtain

There are several ways to enter the Klamath-Siskiyou bioregion. One way is to fly a small plane at low altitude from a city, such as San Francisco, north to Crescent City, California. The plane trip moves across oak savanna woodlands and then mountain ranges covered with forests. Observing from the plane, clear-cuts, patches of forest denuded of timber, are clearly visible. The observer can also see literally thousands of miles of logging roads built on public and private lands in the bioregion.

Visitors driving into the bioregion from the south, from the metropolitan area surrounding San Francisco bay will travel on US101, the only major north-south highway through the bioregion. There is a historic connection between San Francisco and the northcoast of California. For over 150 years, housing in the San Francisco Bay region has been built and rebuilt with old-growth redwood. The first lumber mill was opened in Humboldt County in 1854. Douglas fir and redwood were shipped by sea from Humboldt Bay to San Francisco. After the great earthquake and fire of 1906, San Francisco was rebuilt again with redwood, and during the post-World War II boom years, after 1945, when the suburbs expanded across the hills surrounding San Francisco Bay, redwood was again used. Old-growth redwood is still used to build expensive decks and stairways on homes in and around San Francisco.

Until 1918 there was no direct road linking the northcoast of California with the San Francisco Bay region. Today when the motorist crosses the Golden Gate bridge on US 101, the Redwood Highway, the motorist passes through many suburbs in Marin County on a six-lane freeway. There are only faint indications that this is redwood country. Except for small dedicated groves of ancient redwoods, including Muir Woods National Monument, all the ancient redwoods were logged from Marin County over a century ago.

North of Novato, the freeway enters Sonoma County and passes through vineyards that have been planted over a landscape that was populated only a century ago by California oak savanna. A few groves of oaks are scattered along the soft, undulating hills. Redwoods and oaks coexisted until the nineteenth century along the rivers of this region. Now both oaks and old-growth redwoods are threatened by human activities. Some biologists conclude that as a biotic community, California oak savanna is even more endangered by subdivisions, grazing, conversion to vineyards, logging, and other activities, than old-growth redwoods.

After the freeway crosses the Russian river north of Healdsburg, it rises over several low ridges, passes the Mendocino County seat of Ukiah, crosses more ridges north of Willits, passes through Laytonville. The oak savanna begins to give way to Douglas fir and some pine along the ridges. The highway climbs to the highest pass, less than 1,600 feet elevation, then descends a long ridge into Leggett where California highway 1, coming from the Mendocino coast, joins US 101, the highway reaches the Eel River. Oak savanna gives way to more dense canopy of mixed forest—Douglas fir, madrone, and redwoods.

Following the Eel River, the highway passes the old logging community of Piercy and, crossing the Eel, enters a small grove of redwoods in Richardson Grove State Park. This is a fragment of a once great stand of ancient redwoods that stretched from the Eel across the mountains to the rugged coastline of northern Mendocino County. In Richardson Grove State Park, the interior complexity and integrity of the forest can still be felt. The closed canopy of ancient redwoods protects an understory of shade-tolerant plants including several species of fern.

The visitor has entered into the great Pacific Northwest bioregion by passing through the Redwood Curtain. Summer fog along the coast and winter rains give a different feeling to the landscape. The movement of salmon from the high mountain streams to the vast Pacific and back to their ancestral spawning streams provide an excellent indicator of the flow of energy through this bioregion and thus many consider the salmon the totem animal of the region.

The Eel descends roughly in a northwest direction as it winds through the Avenue of the Giants. With all its tributaries, the Eel is 3,448 miles long. The waters of the Eel River carry more silt than the Yangtze River of China. The Eel River has an annual average load of 5,846 tons of suspended sediment per square mile. In the 1890s, ocean-going vessels could sail up the Eel to Fernbridge. In the 1990s, the Eel was in danger of losing its resident Coho salmon population because of roadbuilding, logging, grazing, overuse of the fisheries, gravel mining, and the introduction of a species of fish that is native to other rivers in California, but not the Eel. Accidently introduced into the Eel in the 1980s, Sacramento squawfish now infest the entire main stem and the lower reaches of all four major tributaries of the Eel.

An extensive restoration plan for salmon and steelhead has been developed by the California Department of Fish and Game. But the combination of two massive floods in 1955 and 1964 and human-caused changes in the whole river system caused a widening the river bed, extensive gravel bars, and shallow sun-warmed waters. Salmon like cooler water, swiftly flowing with shaded deep pools near the river bank.

Appreciating the forces and events that shaped this landscape both during geologic and historic times helps the visitor contemplate the suffering that has occurred to this river and the meaning of the slogan "think like a watershed."

The Eel River is part of the Six Rivers region—which includes the Van Duzen River, Mad River, Redwood Creek, Smith River, and Klamath River.

This is the heart of the redwood region where the "monarchs in the mist," a paleoendemic species, have dwelled for millions of years. Paleoendemics are remnant populations of a species that once had a much wider distribution. This species of redwoods once were thriving in vast regions of the northern hemisphere from Germany to China. The mature, self-sustaining coastal redwood forests that remained at the beginning of the twentieth century had developed a strategy of survival based on using summer fog and warm winter rains to sustain themselves.

Near the entrance to Richardson Grove State Park is the "Grandfather Tree." As with many other such attractions in the Redwood Empire, the history of the tree is marked by historically significant events in Western civilization—the sailing of the Mayflower, the American Revolution, the Civil War. Is it possible that only when we put the life of a tree in the context of our human history, that its great age and endurance has meaning for us? Humans are only a tiny leaf on the Tree of Life, but somehow the history of the Tree of Life has to be interpreted in terms of short time spans of human history.

When the Redwood Highway was opened in 1918, one of the favorite roadside attractions was the drive-thru tree. Entrepreneurs along the new Redwood Highway carved openings through ancient redwoods so that motorists could drive through them and have their picture taken in their vehicle. Perhaps it is a testament to the compassion of redwoods that they stoically endure the humiliation of becoming a drive-thru tree, another tourist attraction.

For some visitors redwoods are a playground, or a scenic attraction that acts as the backdrop for important social activities such as kayaking on the river, hiking, and taking digital photos of each other standing next to giant redwood trees.

The visitor, continuing north on US 101 from Richardson Grove State Park, can take a "scenic alternative" to the freeway and drive the old highway through the Avenue of the Giants. This "beauty strip" of remnant groves of ancient redwoods along the Eel River gives the visitor a sense of the energy stored in these ancient trees. Fire scars from many fires that burned the understory of the forest, illuminate a history fraught with perils—and yet the ancient trees survived—until the advent of industrial logging in the second half of the nineteenth century.

In the context of global warming, trees are considered carbon sinks, that is, they hold CO_2. Therefore preservation of large forests of trees in areas such as Borneo in Indonesia and the Amazon basin, is considered a high priority for reducing the effects of human-caused global warming.

Chris Maser, forest ecologist and teacher, says that forests are the expression of interactions of time, soil, climate—and *place*.[1] This place includes summer fog drifting into the interior valley of the Eel River, welcome relief from the annual summer drought. Deep soils provide nutrients for the redwoods that have relatively shallow root systems. When one of the Monarchs in the Mist falls we sense that a great life is changing form. Fallen giant redwoods at the Founders' Grove, named for the visionaries who established the Save the Redwoods League in 1918, and in Rockefeller Forest continue to sustain the life of many beings including ferns, mushrooms, and a myriad of insects. Because of their age and height, some of these fallen giants were given names such as the Flatiron Tree. Respecting and honoring ancient trees, the elders of their species, is an old and honored tradition in this nation. It is especially important to our psyches, accustomed as we are to instant everything, to appreciate the meaning of sustainability by contemplating the lives of ancient redwoods even when they have fallen.

Stopping in Garberville, the urban hub of southern Humboldt County, the visitor senses both the laid back attitude for which Humboldt County is famous, and the entrepreneurial spirit of the "new settlers movement." After the Summer of Love in San Francisco in 1967, some of the counterculture hippies dropped out of the urban scene. Drifting north along US 101, they discovered cheap land, abandoned logging towns, and long, hot summers in southern Humboldt County. The Flower Children began growing pot, "Humboldt gold": they grew first for their own use, then in larger and larger commercial operations. By the late 1970s the Emerald Triangle—Trinity, Mendocino, and Humboldt counties—had become famous for the quality and quantity of Green Gold coming out of the hill country. A county road crossing the hills east of Garberville crosses many old ranch roads, logging roads, 4-wheel drive tracks that lead past locked gates posted with "no trespassing" signs into the backcountry where the Green Gold is grown.[2]

During the summer and early fall, government agents working for the Campaign Against Marijuana Production (CAMP) fly helicopters across these hills, searching for pot plantations. One grower, who I interviewed repeatedly during one growing season, eluded CAMP by setting up an elaborate pot growing operation on federal land managed by the Bureau of Land Management (BLM). He and his partner, backpacked water storage tanks, hundreds of yards of black plastic drip line, and fertilizer to their site. Through the long, hot summer they camped at their site, watched their plants grow, backpacked more water. Although they planted in plastic bags hung in trees, rodents ate some of their prize buds just before harvest. An early autumn rainstorm damaged other plants. After harvest, when they counted up their sales, they barely broke even. But it was a summer adventure.

The experience of my acquaintance probably was not typical of the growers of Humboldt County. More successful entrepreneurs have literally gone underground with their operations, growing cloned pot plants under artificial light to elude CAMP. In 1998 police discovered one indoor operation in a building that looked like a house complete with children's toys in the yard and flowers growing in the garden. Inside the "house," however, was a sophisticated growing operation with over 9,000 plants, a drying room, generator, and packaging room. The growing sophistication of the Humboldt Green Gold culture has been documented by anthropologists, criminologists, and police agencies. Large laboratories where methamphetamine is "cooked" have also been discovered in the backwoods of Humboldt and Trinity counties. The casual visitor, however, sees only the surface of that culture. After an espresso coffee in Garberville, the casual visitor heads on north on the Redwood Highway.

As the visitor drives north, more chip trucks and log trucks crowd the highway. Log trucks bearing bumper stickers with slogans such as "Forests for the Future," "We plant eight trees for every tree we cut," and "We Care," race each other to the mills on Humboldt Bay. Industrial loggers, however, are planting "super" trees developed at their tree nurseries to grow faster than wild trees. And the implicit message of "we care" is that "we care for the bottom line of our corporation and the wood we supply to humans." This anthropocentric bias is the basic premise of the philosophy of resourcism. Resourcism is based on the premise that Nature is a collection of natural resources that can be exploited by humans with technology and labor. Trees are considered a renewable resource.

Some critics of resourcism argue that old-growth redwoods are not a renewable resource because humans cannot plant a seedling and grow a redwood tree to two hundred years, much less two thousand years.

Logging trucks can carry a single log—part of a tree that might have stood for over a thousand years before it was chain-sawed to provide decking for homes in Marin County, but during the first decade of the twenty-first century, most logging trucks carry small trees that were considered trash in previous decades, but now are milled for 2X4s in highly automated lumber mills.

Visitors to the region frequently compare what they see with what they are used to seeing. In comparison to the brown, parched hills and valleys of central and southern California, the Redwood Empire does appear, to the casual visitor, green and healthy. Some visitors, arriving in Humboldt County via US 101 exclaim "It's so green and lush. So much forest." Local ecologists call that reaction the "green illusion."

Bull Creek provides an object lesson in the importance of moving beyond the "green illusion" into thinking like a watershed. Using a major donation of money from John D. Rockefeller Jr., the Save the Redwoods League acquired stands of old-growth redwoods on the alluvial flats in lower Bull Creek during the 1930s from Pacific Lumber Company (PALCO) and donated them to the state of California to form Humboldt Redwoods State Park. The League was not thinking about the vulnerability of the upper watershed, then dominated by Douglas fir forests mixed with meadows used for ranching. As the ranchlands outside the park boundaries were logged during the housing boom after World War II, the scene was set for horrific erosion both above and within the park. Responding to high prices for Douglas fir in the post-World War II housing surge, ranchers cut logging roads across steep, unstable hillsides. The resulting "natural" disaster was only too predictable. Two so-called 100-year floods—one in 1955 and the next in 1964—brought torrents of water and soil down the clear-cut hillsides into Bull Creek. Many of the "saved" redwoods, trees within the boundaries of the state park, were undercut as the soil was blasted away from their roots by the force of the floodwaters.

After the floods of 1964, the state of California and Save the Redwoods League began acquiring private property in the upper watershed of Bull Creek and embarked on heroic restoration and tree planting projects and began monitoring the massive movement of silt from the denuded upper slopes of the watershed down Bull Creek into the Eel, and eventually into the Pacific Ocean.

North of Bull Creek, just before crossing the Eel again at the Shively Road, US 101 passes Stafford, a small community that was almost wiped out by a massive landslide from an upslope logging operation, during winter storms in January 1997. In 1997, local activists used the slogan "remember Stafford" at demonstrations and public hearings calling attention to the need for stronger regulation of the logging industry. Clear-cutting, roadbuilding, and associated activities in many of the watersheds of the region continue to threaten the integrity of the bioregion.

After crossing the Eel River at Stafford the visitor passes Scotia, home of PALCO. This was the only remaining corporation logging significant stands of ancient redwood forest. For a century Scotia has been a company town, owned, controlled, dominated by PALCO. During the turbulent 1960s, a writer and photographer did a photo essay on Scotia and titled it "life in the peace zone."

Since Charles Hurwitz, a Houston based capitalist, acquired control of PALCO in 1985 using junk bonds to buy the company, life in Scotia, and in the surrounding redwood forests, could be better described as "life in the war zone."[3]

Stopping in Scotia, the visitor can take a tour of the logging mill, visit the logging exhibit, and visit the fisheries restoration exhibit that PALCO built to show how sensitive the corporation is to environmental concerns. The public and state and federal agencies have a different view of PALCO's behavior on the land. In late 1997, the California Department of Forestry refused to renew PALCO's logging permit for 1998. Because of their extensive record of violating provisions of the state forest practices regulations, the company was put on probation.

The Headwaters purchase with funds from the state and federal governments, of approximately 3,900 acres, ended the "timber wars." PALCO continued to log on its remaining approximately 240,000 acres land base and in 2007 declared bankruptcy. The company sold approximately 25,000 acres of its land base, mostly oak savanna, to an out of state land corporation. The future of these lands has not yet been determined.

North of Scotia, the highway crosses the Eel River twice again, around the declining town of Rio Dell, a town extensively damaged in a large 1991 earthquake.

After passing the conservative town of Fortuna, the highway bypasses Fernbridge, the oldest existing bridge across the Eel, opened in 1914 and a survivor of many floods and earthquakes. The road across Fernbridge leads to Ferndale, a small farming community that has retained much of its Victorian era heritage and is also the site of the Humboldt County fairgrounds. The estuary of the Eel once supported vast flocks of migrating birds on the Pacific flyway. The remaining wetlands are still winter resting sites for migrating birds. However, diking of wetlands to form pasture and use of pesticides by ranchers has taken a heavy toll on the integrity of the estuary.

Rising across a ridge at Loleta, the highway descends onto the Humboldt Bay floodplain. On the right are the low buildings on the College of the Redwoods and on the left, the wetlands of south Humboldt Bay Wildlife Refuge.

Humboldt Bay is the largest estuary and bay habitat area between San Francisco Bay and Washington State. The highway passes Fields Landing where raw logs are exported from the Humboldt bioregion to processing plants in Asia. At King Salmon the highway passes Buell Point, site of the Humboldt Bay nuclear reactor, first commercial electric generating nuclear reactor opened along the California coast and the first closed, in 1976.

Humboldt Bay is the "eye" of the northcoast basins region. Exploring our relationship with Humboldt Bay helps to establish a sense of social changes that have surged through this landscape and shaped the landscape over the past century.

On the Bay side of Buell Point is a small plateau where visitors can look west through the entrance of Humboldt Bay, between the north and south jetties, into the oncoming swells from the vast Pacific Ocean. Looking south across the community of King Salmon, the visitor sees South Bay, bounded on the west by the south spit, on the south by Table Bluff and the east by dikes and by US 101 freeway. Looking north into the North Bay the visitor sees low lying islands in the middle of the bay and beyond, out of sight, vast mudflats. And to the east, above the Elk River is the first low ridge of the coastal range—frequently bathed in clouds during the rainy season, dark green, almost transparent in the sunset.

Buell Point is where Humboldt Bay got its name in 1850 when a party from the *Laura Virginia*, the first European boat to enter Humboldt Bay, decided to name this bay after a man who never visited this bioregion. Baron von Humboldt was a German scientist who was renowned for his explorations in South America in the first half of the nineteenth century. The baron was an ecological fieldworker who preferred trekking in the wilderness to sitting in a musty laboratory.

Some historians say that the Indian name for Humboldt Bay is Qual-A-Wa-Loo, although that has never been definitively proven. The Wiyot Indian name was "Wike." Humboldt Bay officially received its name on March 15, 1850, during the rediscovery of the bay from the ocean by an expedition sent from San Francisco, according to a report in *The Humboldt Bay Region*, California State Historical Association, 1929, although others claim it was not named until the *Laura Virginia* entered the Bay on April 14, 1850.

Spanish explorers seventy-five years earlier, in 1775, put the region on the European map when they landed at Tsurai, the mountain by the sea, a Yurok village, twenty miles to the north of Humboldt Bay, after missing the entrance to Humboldt Bay.

During the post-World War II era, when Ike Eisenhower used the slogan "atoms for peace" to defuse some of the fear of nuclear holocaust that was sweeping the nation during the Cold War, PG&E, the giant California-based investor-owned utility, began thinking of a string of nuclear-powered electric generating plants strung along the California coast from Humboldt Bay to Diablo Canyon. Some leading conservationists, including David Brower, initially embraced the idea of nuclear power as an alternative to building more dams on California rivers for hydroelectric generation. But during the 1960s, David Brower, David Pesonen, and other vanguard conservationists realized the long-range impact of placing nuclear-powered generating plants along the California coast—one of the most geologically active coastlines in North America.

The Humboldt nuclear generating facility was started in April 1963. This reactor is very small in comparison to the giant facility PG&E later built at Diablo Canyon. After opening the Humboldt Bay facility, new information became available on the geology under Buell Point, and the facility was closed in 1976. Closing a nuclear facility means that the radioactivity in the facility—in the fuel rods, in the concrete of the structure itself, in the soil surrounding the facility—is still on the site. In 1998 PG&E announced plans to begin decommissioning the plant.

Some concerned critics, including Redwood Alliance, believe that decommissioning will cost more than the original construction of this facility.

Scientists tell us there are many types of radioactive particles, some with short half-life (the time it takes for half of the radioactivity to decay) and others with extremely long half-lives. High level radioactive spent fuel must be removed from the storage pool before decommissioning can occur. The radioactive material in the Humboldt nuclear facility—including the concrete walls of the facility—can be containerized and trucked out of the facility to a waste disposal site, such as the proposed national radioactive waste disposal site under Yucca Mountain, Nevada, but there are hazards involved in transporting wastes from a nuclear reactor. Furthermore only the high level radioactive waste (generally limited to spent or irradiated fuel rods) would go to Yucca Mountain. The rest will go to low level waste facilities.

Another option is on-site containment. Contain the decaying radioactive sources and monitor them and wait a while—maybe a hundred years—while continuing to monitor the site. Redwood Alliance, the grassroots, local coalition that maintains a citizens alert on the nuclear reactor, takes the position that "our" waste should not be foisted on other communities. Since the Humboldt region "enjoyed" the benefits of the plant (its electricity), residents should be willing to live with its waste products. Redwood Alliance is a strong proponent of dry cask storage, the only way to move the fuel from harm's way when the next earthquake happens.

Most lay citizens, including this writer, do not understand all the intricacies of radioactive facility decommissioning. However, we can contemplate our relationship and our responsibility for this artifact of twentieth-century civilization.

Once, as part of a course taught at HSU, I took my students on a fieldtrip to Buell Point to engage in an exercise in contemplation of the fire in the belly of Humboldt Bay. Emerging from our bus, we had agreed to walk silently up the trail, through low growing coastal trees and shrubs to the plateau on the ocean side of the nuclear facility. Barbed wire fences and security systems "protect" the public from the nuclear facility. Once on the plateau, facing the nuclear facility to the east and the entrance to Humboldt Bay on the west, I asked the students to form two circles, those on the inside circle facing outward, those in the outside circle facing inward. The students in the inner circle, facing outward, represented the beings of the future—centuries in the future. They were mute, looking backward in time to those humans who made the decision to build this facility in the second half of the twentieth century. The outer circle represented the present generation. Their job was to speak to the beings of the future and explain to them what happened in the late twentieth century. What made industrial civilization spin out of control, lusting for greater and greater domination over nature, lusting to bring all the "decadent, old-growth forests" into management? Facing the future beings so, some students of the present generation tried to explain the logic of capital expansion, the hubris of men using mega technology, the ideology of economic growth and development, but soon all of them fell silent. Returning to the silence of this place, the silence of radioactive particles going through their decay processes, the silence of knowing what had happened but at a loss of words to express what had happened.[4]

Some of the students in the outer circle, the present generation, began crying, softly crying in the midst of this place with great bellowing clouds blowing across the bay, seagulls and brown pelicans drifting overhead in the afternoon breeze.

Spontaneously, in a beautiful act of compassion, the beings of the future—the students in the inner circle—reached out to touch, to hold the beings of the present, to embrace their suffering.

Silently, our little group of pilgrims retreated from the plateau into a small grove of alders underlaid with ferns, at the bottom of the hill. As humans have for millennia, we found comfort in the midst of a grove of trees. Trees are benign beings, always giving shelter, shade, comfort to those humans who enter the dark woods with respect.

Under the shade of the trees, students spoke—of their despair, their sense of guilt, their sense of faith that somehow we, the present generation, will be able to appreciate the implications of our thoughtless actions.

If industrial civilization can understand the importance of nuclear sites as sacred places, then we can begin the real work, the working of restoring our commitment to the place wherein we dwell. Bringing this nuclear facility into the heart of the community of humans who dwell in the Humboldt Bay region signifies our commitment to future generations of humans and other beings and, more importantly, our commitment to the ground of being out of which the manifest world arises.

Driving further north on US 101, a visitor crosses the Elk River, the only river that drains directly into Humboldt Bay and enters the city of Eureka.

Fort Humboldt State Park, on a bluff above historic wetlands, commemorates the pre-Civil War outpost of the U.S. military that was established to protect European settlers from the Indians but ended protecting local Indians from genocide from European settlers. The massive regional mall, the Bayshore Mall, sits atop historic wetlands—once habitat for bird and marine organisms. This is also the historic site where European settlers held Indians in a concentration camp–like facility in the 1850s, where many of them died of exposure.[5]

Passing through downtown Eureka, the visitor catches glimpses of the huge exhaust stacks of the massive pulp mill across Humboldt Bay at Samoa. A two-block detour from US 101 to Second Street brings the visitor to the historic main street of Eureka. During the 1870s Second Street was crowded with Chinese workers, out of work loggers, prostitutes, and an assortment of merchants servicing the needs of men in the logging and shipping industries. Drugs and sex for sale have been a mainstay of the Eureka waterfront area for over a hundred years.

During the 1980s the politicians in Eureka redeveloped Second Street as Old Town—an area of quaint shops and brick streets with a pseudo-Victorian village atmosphere. At the top of Second Street squats the Carson Mansion, probably the most frequently photographed building in Eureka. Built of exotic hardwoods, some imported by ship from South America, the Carson Mansion was the home of a pioneer capitalist logger, William Carson. For many years the Carson Mansion has been the home of a private club whose members include most of the power elite in Humboldt County.

Humboldt County has been an internal colony of capitalist expansion for 150 years. Why not make it a colony of the globalized economy for the next one hundred and fifty years? The future of Eureka depends on decisions it makes concerning its relationship with Humboldt Bay. Will the Mothers and Fathers of Eureka achieve their goal of deepening the ship channel in Humboldt Bay so that large ships can enter the port and make Eureka an outpost of a globalizing economy?

In the center of Eureka between Fourth and Fifth Street is the largest structure in Eureka—the massive, multistory county jail, the Pink Palace, or Humboldt Correctional Facility as it is officially named. The county jail, the pulp mills, the Carson Mansion and the Humboldt nuclear power plant are icons and symbols of the culture of domination that has held sway in this bioregion for the past 150 years. During the 1960s, soon after the pulp mills at Samoa were opened, an ecologist called Eureka the "armpit of the Northcoast" because of the pervasive smell of rotten eggs that drifted across the city (sometimes extending as far as Arcata, ten miles across the bay). The chlorine-based pulp mills are now closed, but for many residents of Humboldt County, the stench surrounding Eureka still pervades.

In the middle of the bay are platforms holding oysters. Arcata hosts an annual oyster festival as part of its attempt to promote local, small industries, but Japanese oysters are an exotic, introduced species in Humboldt Bay and expansion of existing oyster farms may have increased negative impacts on Humboldt Bay. Native bat rays prey on oysters and oyster farmers have been known to kill bat rays.

There is a native oyster, the Olympic oyster (*Ostrea lurida*), with remnant populations in Humboldt Bay. To what extent it was present historically is unknown, but its potential habitat has certainly been reduced and what presently remains is mostly occupied by the farmed alien, the Pacific or Japanese oyster (*Crassostrea gigas*). The Olympic oyster does not fit the economic preferences of oyster farmers. It is slow growing and small, although connoisseurs rate its flavor highly.

One island in Humboldt Bay, Gunther Island, was the site of a mass killing of Wiyot Indians on February 26, 1860, when the Mothers and Fathers of Eureka sent armed militia to the island to murder Wiyot Indians. The massacre was the beginning of a systematic slaughter of Indians in villages dwelling along the lower Eel River and near Humboldt Bay. During 2007, the Wiyot tribe announced plans to restore Gunther Island, remove toxic wastes including dioxin, from the island and to host the World Renewal Ceremony again.

Sacred places are part of bioregional identity. Ceremonies at sacred places bring the community together to renew commitments to each other and to the place wherein they dwell.

Humboldt Bay since 1854 has been the primary export harbor for wood logged in the region. When the two pulp mills on the Samoa side of the Bay were opened in 1964, the great forests were reduced to mere chips for making pulp in chlorine-based mills. The Humboldt Municipal Water District agreed to provide millions of gallons of water a day to the pulp mills via a pipeline from the Mad River to the site of the pulp mills. A new dam was constructed at Ruth, on the upper Mad River, to regulate the flow of the river so that sufficient water would be available even during the summer months.

The opening of the two pulp mills was celebrated by the Mothers and Fathers of Eureka, but the smell of rotten eggs that drifted with the winds as far as Arcata told a different story. Air pollution and water pollution during the 1960s were considered necessary externalities of industrial development. By the 1980s, however, after the passage of federal legislation to regulate and reduce air and water pollution, the pulp mills were considered blights on the landscape. However, it took a small group of surfers to make waves with the giant logging corporations and bring those corporations to their knees in a landmark settlement. Surfing lawyer Mark Massara, attorney for Surfrider Foundation, brought a suit against the owners of the pulp mills after documenting that the two pulp mills, which dumped 40 million of gallons of untreated waste into the ocean each day, were flagrantly violating clean water standards set under the Federal Clean Water Act.

In 1991, in an out-of-court settlement, the two pulp mill operators agreed to pay nearly $5.8 million in fines and stop polluting the ocean off Humboldt County. The pulp mill owners agreed to spend more than $50 million to dramatically reduce toxic discharges.

By the end of the 1990s, the two pulp mills were relics—expensive, contaminated relics, but relics, nevertheless, of the most aggressive phase of industrial capitalist expansion in Humboldt County. The owners of one of the pulp mills closed their mill rather than invest in pollution abatement equipment mandated under the out-of-court settlement with the EPA and Surfrider Foundation. The owner of the other pulp mill, Louisiana-Pacific, retrofitted their mill with oxygen-based processes then decided to sell the mill rather than continue to operate any pulp export facility on Humboldt Bay. The pulp mill was then sold to a Chinese corporation and renamed Evergreen.

The history of the pulp mills illustrates one facet in the restructuring of the logging industry throughout the bioregion. As old-growth forests were depleted, the industry turned to "industrial forestry"—growing trees until they were twenty or thirty years old and cutting them for pulp or fiberboard. The industry then engaged in aggressive restructuring and mechanization. Older, inefficient mills were closed and larger, mechanized mills were opened. Millworkers were replaced by machines. Louisiana-Pacific went further. They replaced American workers with cheaper labor in Mexico. They built a new processing mill in Baja and shipped raw logs from their dock on Humboldt Bay to Mexico for processing and reshipment to the southern California market.

When Louisiana-Pacific sold all their timberlands in Humboldt County to Simpson Timber in 1998, only three large timber holding corporations remained in Humboldt, Del Norte, Siskiyou, and Trinity counties—Simpson (now called Green Diamond Corporation), Sierra Pacific, and Maxxam (PALCO).

Slipping into the dark waters of Humboldt Bay, exotic invasive alien species have impacted the native plant and animal communities of Humboldt Bay and other areas of the bioregion. Some of these invaders come aboard log ships from Pacific Rim ports—yes, ships bringing logs to Humboldt Bay. In a globalizing economy, enterprising capitalists ship plantation grown pine from Pacific Rim nations and logs from the emerging logging industry in the Russian Far East, into American ports, including Humboldt Bay. Although the wood cargo is sprayed with a pesticide (methal bromide) while at sea, some scientists suggest that even a few fertile Asian gypsy moths (*Lymantria dispar*) entering the forest system of the Pacific Northwest by escaping from infested logs could infect and destroy a billion Douglas fir trees within a few decades depending on the nature of the incipient invasion and how long it takes the invasive species to establish itself.

The Humboldt Bay ecosystem itself is threatened with the possible invasion of European green crab (*Carcinus maenas*). This voracious predator can quickly invade and impact intertidal marine ecosystems such as Humboldt Bay. Shorebird foraging is only one factor that could be affected by this invasion. European green crabs already have invaded and impacted other major intertidal marine ecosystems in California. It invaded San Francisco Bay and probably found its way northward in the 1980s via surface currents.

While scientists have been documenting the impacts of exotic invasive aliens for many years, only since the 1990s have conservationists begun to understand the importance of political activism and public education on the effects of exotic invasive alien species and advocate for more coordinated prevention and management goals.

Exotic invasive aliens are only one of the threats to the integrity of the Humboldt Bay ecosystem. Plant and animal invasions are quiet invasions that can have long-term effects. They don't attract much news coverage or attention until the invasion is irreversible. Oil spills are different. They are dramatic events that can have long-term effects on an ecosystem. The oil spill which took only twenty minutes to occur in November 1997 in Humboldt Bay is an example of this type of human intervention.

On November 7, 1997, nearly 4,700 gallons of "Bunker C" fuel oil leaked into Humboldt Bay at the Louisiana-Pacific dock during the early morning. The *Kuri*, a wood chip ship, was being repositioned at the dock and hit the steel and concrete dock rupturing the hull at the fuel tank. Within half an hour all the contents of the fuel tank had leaked into Humboldt Bay. Twenty-four hours later tarlike oil had spread out of the entrance channel and was spreading north toward the oyster beds and south toward Humboldt Bay Wildlife Refuge.

Even as hundreds of workers were siting booms, shoveling oily sand into white plastic bags, and caring for stricken wildlife, the mainstream business leaders of Eureka were supporting dredging the ship channel to greater depth so that larger ships can reach the docks of Eureka and Samoa. The oil spill emergency response team worked twenty-four hours a day to contain the spill and assist damaged birds. Loons, sea gulls, and white-winged scoters were found covered with oil. Many were dead. Those that were living were transferred to the Marine Wildlife Care Center at HSU. The birds were washed with a solution of Dawn detergent and water and toweled off. Most birds remained at the center for two to three weeks.

Under the lead of the U.S. Coast Guard, personnel from California Fish and Game Department and U.S. Fish and Wildlife Service as well as contract crews and volunteers were mobilized to clean up the spilled oil and care for oil-covered birds. Officials with the Oiled Wildlife Care Network within the Department of Fish and Game directed rehabilitation efforts at the HSU facility. These officials worked with faculty, students, and community volunteers. Biologists say that the long-term consequences of this oil spill will not be known for some time.

For 150 years the Mothers and Fathers of Eureka have been playing Russian roulette with the fragile Humboldt Bay ecosystem. In their desire for economic development, for money and power, they installed more and more industrial technology on this fragile, sentient being called Humboldt Bay. As a result, the Mothers and Fathers of Eureka have damaged the interrelations between ocean, tidal flats, wildlife and habitat.

Emergency response workers delicately holding a common murre insert a long tube into its stomach in an attempt to rehydrate the bird. Birds in these oily waters ingest toxins when they try to preen their feathers or to feed. They must be rehydrated before the oil can be cleaned off their feathers.

Humboldt Bay has experienced previous oil spills. Perhaps the largest oil spill occurred on November 24, 1968, when a Standard Oil pipeline at Bucksport, in Eureka, ruptured, spilling 81,680 gallons of diesel into Humboldt Bay (Eureka *Times-Standard* November 16, 1997). The response to that oil spill seems to have been minimal. No impact assessment was done by state agencies and the North Coast Regional Water Quality Board did not even discuss the oil spill at its next meeting.

That was in 1968. At that time the only conservation groups active in the county were the Audubon Society and the Sierra Club, and the Sierra Club was preoccupied with the dedication of Redwood National Park. In 1968 there was no NEPA, no ESA, no EPA, no state Coastal Commission and no national policy to protect wetlands. The Mothers and Fathers of Eureka were logging old-growth forests regardless of the environmental harm.

The contrast between the response to the oil spill in 1968 and November 1997 shows the development of legislation and institutions that incorporate some aspects of an environmental ethic. People from diverse occupations and life situations expressed concern for the suffering of Humboldt Bay during the November 1997 oil spill.

What are the ecological and social implications of a large oil spill? Why are woodchips and wood pulp, not sawed timber, not any processed wood products leaving Humboldt Bay in large, foreign owned vessels? Woodchips, chipped in the woods by a giant machine, are trucked to Samoa in huge trucks, and dumped onto ships bound for the other side of the Pacific. The forests of Humboldt County have suffered many abuses over the past 150 years, but nothing as demeaning as this—chipped and shipped away. Is the Klamath-Siskiyou region nothing but a chip factory in the minds of corporate owners?

Georgia-Pacific Corporation (GP) bought Hammond Lumber Company, including the logging town of Crannell in Little River watershed, during the 1950s. GP (before it split into GP and LP) built the pulp mill at Samoa in the early 1960s. By their own admission, LP violated their own sustained-yield forestry plans to sustain production at the pulp mill and particle board factory and sawmill. LP deployed aircraft to engage in spraying of herbicides on Little River watershed, endangering the water supply of Trinidad and Westhaven. In the autumn of 1997, LP announced it was selling $1 billion worth of assets in northwestern California, including its pulp mill and 300,000 acres of mostly redwood forests. Simpson Timber Corporation bought all LP's timberlands in Humboldt County in 1998 and transferred the land to Green Diamond Corporation.

Are forests and watersheds only corporate assets? Is the forest only a tree plantation? Is the forest just a chip factory? For 150 years Humboldt County has been treated as a colony of industrial civilization and the corporations have treated the complex forests as a factory merely to produce profits for the shareholders.

Humboldt Bay is a living being, a vibrant breathing and moving entity, sentient, alive.

Great earthquakes shake the shores of Humboldt Bay. Storms sweep across the Pacific Ocean bringing monsoon-like rains. Windstorms blow down whole groves of trees. Forest fires sweep through tinder dry mountain regions during the summer drought, but no natural event spills diesel fuel across the waters of Humboldt Bay.

NOTES

1. Ecoforester, Chris Maser (1989, 1992, 1994), has documented the natural history of an ancient forest and the intricate harmony of the interactions between the varied living organisms, natural processes and the existing landscape in the forest ecosystem.
2. The growing of marijuana in the southern part of Humboldt County was one of the activities undertaken by some of the members of the back-to-the land initiatives who migrated from the Bay Area to the northcoast of California. For an account of this, see Ray Raphael (1985).
3. For a detailed account of the takeover of the Pacific Lumber Company by Maxxam Inc. and the various protests and conflicts that resulted, see David Harris (1995), The Last Stand.
4. For experiential education through workshops and lectures to promote and realize bioregional themes and concepts. See, for example, the writings of Joanna Macy and John Seed (Seed and Macy 1988; Macy 1991).
5. For a historical study of Native American genocide in Humboldt County, see Norton (1979)

Chapter 3

The Energy Dilemma

When the nuclear electric generating plant on Humboldt Bay was in operation, the electric power company Pacific Gas & Electric had a large sign near the plant announcing that the Humboldt Bay Nuclear Reactor was at the service of consumers.[1] When the plant was closed in 1976, the sign was removed. However, the radioactive rods that fueled the electric generator remain in the plant.

The history of nuclear generating power plants illustrates the dilemma of energy during an age of affluence.

For most of the history of human habitation of the Klamath-Siskiyou region, wood was the basic form of energy used for cooking and heating. Even in 2007, many homes in Humboldt, Del Norte, and Trinity counties have wood burning stoves or some divertive of wood such as wood pellet stoves. Currently, however, wood pellets are shipped to the region from plants in the state of Washington, British Columbia, and Ohio. Of course, it takes fossil fuel to ship wood pellets via trucks from distant locations into Humboldt County.

Electricity is distributed via power lines stretching from the Central Valley of California to the coast. Natural gas is also transported via a pipeline from the Central Valley of California. The natural gas shipped into Humboldt County originates in Alberta, Canada.

Gasoline is shipped to local gasoline stations via trucks from refineries on San Francisco bay and from other refineries on the Gulf Coast of America.

The uncertain state of the electric system is illustrated almost every winter in the Klamath-Siskiyou region when a wind storm knocks trees across power lines. Thousands of consumers are without electricity for a week.

Some people search for "alternative" energy sources. Tests are conducted on the use of wave energy to generate electricity. Windmills are suggested to generate electricity. Solar panels are sold to consumers who want to use sun power to feed electricity into the central electric distribution system or store solar generated electricity in batteries.

Some energy efficiencies in buildings have been achieved due to state legislation passed during the 1980s that required insulation in housing and commercial buildings. Energy efficiency, however, does not end energy dependency.

Gasoline prices increase, and even if ethanol is added to gasoline the corn that is used to produce ethanol is grown in Iowa, not in the Klamath-Siskiyou region.

Solar power, wind power, and wave power are intermittent and cannot be stored for long periods of time. People who proclaim they are "off the grid" have batteries, but during months of cloudy weather, or months of low wind, their ability to store electricity is extremely limited and not at all available for commercial buildings or for a large facility such as the campus of HSU. Some people who claim to be "off the grid" use gasoline-fueled electric generators.

The global implications of disruption of oil supplies from the Middle East is the subject of nightly news reports on television. Even a minor typhoon moving across the Persian Gulf can cause oil prices to surge on world markets.

The county bus system in Humboldt County proclaims they are using bio-diesel. Many scientists ponder new technologies such as hydrogen fuel cells, but commercial vehicles using hydrogen fuel cells are not on the market. Even increasing miles per gallon of gasoline in commercial vehicles is difficult, according to Detroit automakers.

The energy future for the Klamath-Siskiyou region is dependent on decisions made by corporations and governments over which local residents have little control.

Conservation is a limited option. Housing can be retrofitted with energy efficient insulation. Residents can use lower powered light bulbs. Higher density housing would concentrate the population thereby reducing the number of miles driven for daily chores. However, residents remain dependent on electricity and gasoline.

Proposals have been made to dock huge ships carrying Liquefied Natural Gas from natural gas fields off the coasts of Australia and Indonesia in Humboldt Bay. The gas would then be reheated and transported via pipeline to metropolitan regions of California.

Wood burning stoves appear to have an advantage since wood is abundant in the region. However, the labor costs of cutting, stacking, and moving wood are more expensive than its current value as a source of energy. Wood for the most part is used by visitors to local state parks. Wood is sold for $10 a bundle to visitors who burn the wood in a designated fire ring in camping areas of state and county parks. The ancient practice of singing and dancing around the campfire as a basic human social action is currently reduced to campfire in a designated fire pit in a state park.

The developing culture of bioregionalism has opportunities to bring forth alternative energy sources. The current generation of corporations and social activists consider the challenge of alternative energy sources to be one of their highest priorities. The Standard Oil Corporation, Chevron, advertises they are based on "people power." The international oil giant British Petroleum (BP) has taken as its slogan "Beyond Petroleum" and has endowed an energy research institute at University of California, Berkeley, with $500 million to research alternative energy sources.

For thousands of years, humans have sat around campfires, talking, dancing, and singing. During the era of global warming, future communities may be much like communities in the past, with small populations relying on wood as their primary source of energy.

There is a low probability, however, that current food production within the bioregion can be maintained and developed, without gasoline, ethanol, hybrid vehicles, or other power vehicles.

Currently over 90 percent of the food consumed in this bioregion is trucked into the region along US101 and California 299.

Arable, agricultural land on the upper Trinity River, in Trinity County, is drowned under the waters of the Trinity Dam. Water from this reservoir is piped across the mountains to the Central Valley of California where it is used for agriculture.

Water diversions on the upper Klamath River, in Oregon, are used to irritate low value crops such as potatoes and alfalfa. These crops are sent to markets beyond this bioregion. Crops grown in the upper Klamath River watershed cannot be considered "local" produce on the coast of Humboldt County.

With rugged, mountainous terrain, the primary agricultural lands are along rivers and in the deltas of rivers. There are ranches and dairy farms in the region, and beef from some of these ranches is advertised as "homegrown" in local grocery stores.

A century ago, many people, even in cities and towns, kept chickens and sheep in their backyard. "Homegrown" then meant that people grew vegetables and fruit trees and canned them at harvest time in their own kitchens, using wood burning stoves for use during the winter months.

Residents of the bioregion must learn old skills in order for the communities within the bioregion to become more self-sufficient.

More residents can take up sustainable hunting. Deer and elk are available. Wild game is sold only in a few grocery stores in the bioregion. Looking to the past as a way to the future requires massive changes in current lifestyles but is compatible with the overall statement "rich in ends, simple in means."

This bioregion does have local breweries, although hops to make the beer is imported from outside the bioregion. Locally made beer sold at the breweries and bottled and sold in local taverns and stores is both part of the diet of local residents and when drunk in cafes is part of local community solidarity.

There are wineries in Mendocino County and Humboldt County and local wines are part of bioregional culture. The art of winemaking based on local terrain is another old tradition that can be part of future bioregional culture.

NOTE

1. The impending scarcity of fossil fuel and the various scenarios that could result are discussed by James Kuntsler (2005). In terms of climate change and the energy scarcity that can result are examined by Jorg Friedrichs (2013). In terms of the energy scarcity scenario, and what can be done and should be planned in a practical fashion as Bill Devall addresses in this chapter, see Richard Heinberg (2005).

Chapter 4

An Enlightened City? Arcata, California

Resting on the Plaza in Arcata, under the gaze of the statue of President McKinley, surrounded by aging panhandlers and wannabe post-hippies, college students, casual shoppers, and historical buildings—including the Jacoby Storehouse on the southwest corner of the Plaza where mule trains were formed in the nineteenth century to take supplies to the mountainous outback regions—a visitor can begin orienting to the Arcata scene. But don't light up a cigarette or walk your dog on the Plaza or ride a skateboard. In the politically correct 1990s, the "progressive" Green Party city council of Arcata cracked down on many of the small pleasures of life. By 2007, even "loitering" on the Plaza is considered an act of crime.

Still, in 1997, the editors of the *Utne Reader* called Arcata one of the ten "most enlightened cities" in America. The article in *Utne Reader* began "Got dreadlocks? Got a VW bus? Got a cause? If so, you've got a home in Arcata." Other journals and newspapers followed suit with purple prose about the cool life in Arcata appearing in the *Wall Street Journal*, *Los Angeles Times*, *Sierra* magazine, and *People* magazine. *USA TODAY* began with a question, "How green is Arcata? As green as the redwood forest that surrounds it." The so-called "green" party that had a 3-2 majority on the city council for two years never voted as a bloc and most of the media hype reported on actions by the city council long before the "greenies" were elected to office.

Green bikes are free bikes on the Plaza. Pick up a green bike and ride it around town. Return it somewhere near the Plaza, but don't put a lock on it. Locks will be cut off. Bike lanes are painted on several of the streets through downtown Arcata. The city council, with a Green Party majority, was diligent in advocating bikes in Arcata. To ride a skateboard go to the designated skateboard park at the end of H Street—and be sure to wear a helmet (provided free) and kneepads or you will be fined (the Green Arcata city council is pleased to provide law and order even for cool skateboarders).

Arcata was a logging town in 1968 when I arrived to live in this bioregion. Male students in classes at what was then called Humboldt State College wore calkboots to class and pulled the greenchain on swing shifts and during the summer break from school. A tepee burner was still burning at a mill in Arcata. Pulling the greenchain was hard, manual work of separating wood of different lengths and sizes coming off the saw and stacking them for shipment. After working in the woods or in the mills, men were ready to party on the weekends.

All of the logging mils in the Klamath-Siskiyou bioregion were automated during the 1980s. Those that were not automated by their owners were closed. In 2007 there are no logging mills in Del Norte County and only a small mill in Trinity County, in Weaverville.

During the late 1960s the Arcata Plaza was the focus of a thriving working-class culture—lined with bars on the north and south side. Some of the same bars still remained on the north side of the Plaza in the 2007 including Toby and Jacks and the Alibi. The bar scene in 2007, however, is a pale remnant in comparison to the 1960s. On Saturday nights during the 1960s the dance bars were crowded with young loggers and college students. The pool tables were active. High stakes games were played after midnight on those tables.

Beginning in the early 1970s, schools for Yoga, Aikido, and other martial arts were organized in Arcata. The Arcata Zen Group has held regular zazen and sesshins for two decades in Arcata. While protests on the Plaza and protesters in the ancient redwood forests have grabbed headlines, direct action also includes spiritual practice. "Why do you practice zen?" I asked a student at the Arcata Zen Group. "I sit for the enlightenment of the myriad of beings," he replied. That sense of faith, commitment to practice, and commitment to the Sangha, the community of all beings, enhances the well-being of Arcata.

The greening of Arcata began in the late 1960s, during the intense battles over establishing a Redwood National Park. The Sierra Club established a group in Arcata to show national politicians that there was local support for the proposed national park.

In April 1970, a national teach-in on the environment, organized by young activists inspired by Senator Gaylord Nelson of Wisconsin, was held simultaneously on many college campuses, as well as in Washington, DC, On the HSU campus, students fired with enthusiasm for wilderness organized their own Earthday and presented the first street theater, teach-in, and pro-wilderness demonstration in Humboldt County history. Charlie Bloom, a university librarian and avid backpacker, helped students interested in backpacking and mountaineering form the Boot n' Blister Club. College students began exploring the rugged backcountry of the Trinity Alps and the Siskiyou mountains. Many of these students became active in the fights for wilderness during the 1970s and in the campaign for expansion of the Redwood National Park. For example, the photos of logging the Redwood Creek watershed taken by student activist Dave Van de Mark are now archived in the Redwood National Park historical collection. From the enthusiasm of Earthday 1970, students and local residents organized the Arcata Recycling Center and the Northcoast Environmental Center.

Arcata was the first city in the bioregion to oppose freeway expansion. In 1971, concerned citizens formed the Stop at Four Committee to oppose massive expansion of the US 101 freeway cutting Arcata in half. The city council favored a plan for a six-lane freeway, but the Stop at Four Committee, reacting to the first wave of revolt against freeway expansion that was sweeping across the nation, called for fewer lanes on the new freeway through Arcata. The committee got some design modifications from the California Department of Transportation (CalTrans), but more importantly they got some of their activist leaders elected to the Arcata city council replacing the good old boys who had ruled Arcata for generations.

The reformers began making changes in the way the City of Arcata treated its sewage, its creeks, and its farmlands and forests. Wetlands were protected and urban sprawl was contained. Growth for growth's sake was no longer acceptable in justifying new projects. The reformers espoused a philosophy of "limits to growth," a philosophy popular during the early 1970s and associated with such writers as E. F. Schumacher, author of *Small Is Beautiful*. Grassroots conservationists in Arcata wanted to limit population growth in Arcata, protect historic wetlands and forests near the city, and protect the open space along the US 101 corridor between Arcata and Eureka.

Arcata is a university town, and some university professors provided ideas, and leadership, for green reform in Arcata. The Arcata marsh waste-water system was designed by university professors. University professors and former university students served on the city council. Environmental engineering, environmental ethics, and natural resources management are well established programs at HSU. Other professors, however, were closely identified with the logging industry and were appointed by Republican governors to key positions on the State Board of Forestry and California Department of Forestry during some of the most bitter battles over logging during the 1970s and 1980s. In those positions they opposed most of the forestry reforms proposed during the 1980s.

A visitor to Arcata walking from the Plaza south on H street comes to the Arcata Marsh Interpretative Center. George Allen, then a fisheries professor at HSU, and Bob Gearhart, then a professor of environmental engineering, designed the marsh units for sewage treatment. While most other cities enclose their sewage treatment facilities within high fences, the Arcata Marsh

Sanctuary and sewage enhancement ponds are open to the public. And the Arcata Marsh Sanctuary has attracted both increasing numbers and varieties of birds as well as river otters and humans. People visit the marsh in all kinds of weather. Joggers, walkers, and birders can be found on the trails during rainstorms and sunny days. Hospice uses the Marsh Sanctuary for quiet activities for people in the dying process. Wedding and memorial services have been held at the marsh. During the first week of August each year, a Hiroshima memorial service is held at the marsh. Participants light candles placed in paper boats and sail them out onto the waters of the largest pond in the sanctuary in memory of those who died in the first use of an atomic weapon on a civilian population and to rekindle commitment to a world without nuclear weapons.

Trout placed in Klopp Lake feeds resident river otters, osprey, and fishermen who fish at the lake. Birders flock to the marsh during all seasons of the year to search for resident, migrant, and vagrant birds. Gearheart Marsh, the most open of the marshes in the Arcata Marsh Sanctuary, attracts Northern pintails, mallards, northern shovelers, ring-necked ducks, and American wigeons. Blue-winged teal are found during spring migration.

American bittern, black-crowned night-heron, great blue heron, and great and snowy egrets are all common in the marsh. The marsh also hosts the largest concentrations of wintering Avocets on the North Coast.

The popular success of the Arcata Marsh Sanctuary shows the transformation of consciousness that has occurred over the past two decades concerning marshes. Instead of seeing these tidal wetlands as wasteland or swamps that should be "reclaimed" for productive agricultural or industrial uses, we now see wetlands supporting populations of plants, birds, and mammals. Wetlands are the lungs of Humboldt Bay.

Of course the progressive changes in the culture of Arcata have not occurred without social conflict. After the wave of popular support at the beginning of the 1970s, the endurance of the conservation ethic was sorely tested in Arcata during the late 1970s, over the issue of protection of old-growth forests.

During the bitter controversy over the expansion of Redwood National Park, between 1974 and 1978, there were many demonstrations, both for and against park expansion, held in Eureka and Arcata. Many who supported park expansion were called "greenie faggots" as they walked the streets near the Plaza. Bumper stickers appeared on pickup trucks cruising through town that said "Don't Park My Job" and "Arcata out of Humboldt County." Some forestry students spat on "hippie environmentalists" as they walked across the Plaza. A forestry professor at HSU, Rudi Becking, was hanged in effigy in the streets of Arcata by angry loggers because he advocated reform of California's archaic forest practices law.

After Congress passed the Redwood Park Expansion Act in 1978, the focus of protest shifted to new issues using new methods of activism. Dave Foreman and Johnny Sagebrush, two of the five cofounders of Earth First!, came to the Plaza in September 1981 on their first national roadshow for Earth First!. Singing songs from *The Little Green Song Book* and telling stories of the new, direct action approach to conservation, they began to mobilize a new generation of forest activists. Since 1981 Earth First! activists have been on the frontlines in Humboldt County in the battles to protect old-growth forests.[1]

During the first decade of the twenty-first century, Arcata became a city where housing prices are increasing, the birth rate is increasing, and the master plan of HSU states the goal of increasing the student body from 7,000 to nearly 16,000. A building boom on campus required relocation of the house occupied by the Center for Appropriate Technology to make way for a large office building for professors. The Plaza was redeveloped with boutique shops for clothes, furniture, and bookstores. The city council continues to attempt to remove transients from the Plaza. The president of HSU said that parents of prospective students don't like to see homeless people hanging out on the Plaza.

Homeless people, or people who are homeless, are people who consume less electricity, gasoline, and other resources than the average middle-class consumer in the bioregion. People who are homeless ride bicycles, walk, or use public transit to move from one location to another. Many of them sleep in tents or, during the rainy season, stay at night in a faith-based shelter, such as the Eureka Rescue Mission. People who are homeless are frequently targeted by the police and given citations for being drunk in a public place, possession of a controlled substance, or urination in a public place.

People who are homeless rarely collect together to protest against government.

The only "protests" on the Arcata Plaza during the first decade of the twenty-first century are "silent vigils" protesting the War in Iraq.

The most popular events on the Plaza are family events including Saturday market, annual oyster festival, and the Kinetic Sculpture Race. All of these draw a middle-class audience.

The Kinetic Sculpture Race, held over the Memorial Day weekend, was created by Ferndale-based sculptor Hobart Brown. While some of the rules have changed over the years, the basic rules have remained the same—each vehicle must be pedal-powered, must have flotation devices so it can cross Humboldt Bay and the Eel River, and must be designed as a moving sculpture. The three-day race starts on the Arcata Plaza, pedals to Manila and across the Manila Dunes, crosses the bridge to Eureka, down the freeway to Fields Landing where the vehicles cross Humboldt Bay, then down the South

Spit to Table Bluff, crosses the Eel and the Ferndale bottoms into the city of Ferndale.

The creative ideas for people-powered vehicles could be applied to some form of local transportation. When gasoline becomes too expensive for lower income residents, people-powered vehicles become a viable option.

The Kinetic Sculpture Race occurs only one weekend a year, but each Saturday morning from May through October the Arcata Plaza is ringed with small stalls during Saturday market making farmer-to-consumer connections. The names used by these small entrepreneurs illustrate the variety of local produce and places—for example, Bigfoot Plants, Blue Heron Trout Farm, Carlotta Flowers, Earth N Hands Farm, Earth and Sky Garlic. More than seventy small farms from all over Humboldt County participate in the market. The diversity of produce for sale at the market reflects the diversity of microclimates in Humboldt County, which nurtures both warm and cool season crops including chard, onions, garlic, lettuce, peas, tomatoes, melons, corn, strawberries, and mushrooms.

Surf's-Up farm brings lettuce, mixed salad greens, squash, and cucumbers to the Arcata Plaza—on bicycles. Affiliated with Alliance for a Paving Moratorium, Surf's-Up demonstrates that there is a feasible alternative to using fossil fuel for gardening. When I interviewed one of the young men who work with Surf's-Up, he commented, "We're showing people that it is feasible to transport garden crops to the market without using fossil fuel. People really appreciate buying from growers who do not use fossil fuel."

Local producers who sell their flowers, vegetables, mushrooms, and herbs and fruit to the Saturday market have also been generous donors of their produce to the AIDS Food Project and Food Not Bombs. Boxes of donated food are distributed each Saturday afternoon to AIDS sufferers who have low incomes and whose illness prevents them from shopping at the Farmer's Market.

The Food Not Bombs collective began providing free suppers on the Plaza to homeless people and wanderers to make a political statement about homelessness. The City of Arcata took Food Not Bombs to court demanding that they obtain health department permits. After several years of wrangling, Food Not Bombs became a legally organized social service in 1998, recognized by the City of Arcata. Some of the old anarchists who established the Food Not Bombs collective complain that not much of the spirit of anarchism remains in the group.

It's hard to be an anarchist or use the rhetoric of "liberation" and "resistance" when the causes which not long ago were "radical fringe" causes have now been established as public policy and administered by bureaucratic public agencies or by private religious groups supported by President Bush's "faith-based" initiative which allocates federal funds to religious groups providing social services.

In 1998 the city council passed an ordinance allowing medical use of marijuana in Arcata based on provisions in Proposition 215, the statewide California initiative in which voters approved medical uses of marijuana in November 1996. The patient can obtain a "medical marijuana" card from the county public health department.

Concern over smoking tobacco has replaced concern over smoking pot. Drinking alcohol is unfashionable. Teenagers wear tattoos and nose rings and baggy pants. Gays are welcome if they conform to "good values" including "domestic partnerships" and "family relations." Bare breasts are not cool on the Plaza. "Make love, not war" is not a fashionable slogan on the Plaza. During the postfeminist era, women entrepreneurs, who are realizing themselves by making money, are in. Skateboards or unleashed dogs are out, but baby carriages are in. Headshops are out, however boutiques and herb stores are in. The Arcata Co-Op is in, but it is not a social movement. It competes with Safeway in the sale of "organic" food.

From 1970 through the 1990s, the progressive culture of Arcata contributed a kind of ambience that encouraged creativity and innovation. For example, Arcata citizens support an educational farm in a section of Arcata called Bayside where anyone can learn how to raise some of their own food and how to share that food with others. Immigrant Hmong residents of Humboldt County share space on the property with the educational farm. Uprooted from their homelands in Laos as a result of their support for the CIA and U.S. military during the brutal War in Vietnam, Hmong families were resettled by the U.S. government in several rural locations in California, including Humboldt County. Accustomed to farming small gardens in their homeland, they applied their skills and knowledge to the microclimate conditions on the Northcoast. Gardening provided a kind of cultural sustenance as well as physical sustenance.

The educational farm borders a small creek. Arcata is built over several small coastal creeks draining into Humboldt Bay—Janes Creek, Jolly Giant Creek, Campbell Creek. These were salmon-spawning creeks before they were logged and contained in concrete-lined storm sewers in many parts of their lower reaches. Awareness of the life of these creeks provides residents of Arcata with the opportunity to practice mindfulness, incorporating the life of the streams into their sense of place. As part of an environmental education project, students from HSU and the City of Arcata stenciled signs over grills on streets draining into each of these creeks reminding residents that these storm sewers are still part of a living creek system.

Walking up G street, northbound, from the Plaza, to 14th street, turning right across the freeway, past the HSU campus to Union street, one can enter the dark woods of the Arcata community forest on a path along Campbell creek. This forest was first logged with oxen at the end of the nineteenth century, then logged again with steam engine skidders. Portions of Campbell creek watershed were acquired by the City of Arcata early in the twentieth century as a water source for fighting fires. Stumps of old-growth redwoods along the trails bear the marks of nineteenth-century logging techniques including holes chopped into the trunk sometimes 15 feet up the butt where planks was inserted to form a platform for the sawyers.

Old-growth redwoods are not a renewable resource. No forester has ever grown a 2,000-year-old redwood or even a 500-year-old redwood or a 200-year-old redwood tree.

Logging old-growth redwoods creates a permanent change in the landscape. Some forest ecologists argue that redwoods do not even begin to develop characteristics of what they call "late seral forests" until they are 150 years old—or older.

The City of Arcata in 1997 obtained certification that the Arcata community forest will be managed based on principles of "sustainable forestry." Although there is considerable disagreement among forest activists on the meaning of "sustainable forestry," the city obtained its "green sticker" certification through Smartwood, a consulting organization that certifies lumber cut to its "sustainable forestry" specifications.

It is ironic that one of the greatest threats to the Arcata community forest is recreation. People are loving the forest to death. As more and more visitors camp in the forest (illegal but widely practiced), trample through the underbrush, ride mountain bikes on the forest trails, and generally enjoy the forest, the integrity of parts of the forest is threatened. Visitors trample the soil around redwood trees thus impacting the roots of the trees. Feral cats, an exotic species in the bioregion, migrate from neighborhoods near the forest and attack some of the birds. Even Earth First! has had some impact on the forest. At a November 1997 seminar on tree-climbing, held in the community forest, about fifteen people scarred a few redwood trees learning how to climb them. As the Earth First! teacher told her students, "There are three reasons for climbing trees—recreation, research, resistance. We are here to resist. We're learning tree-climbing so we can save redwoods from logging. We ask that the trees forgive us today."

Generations of Earth First! activists learned their forest protection skills by practicing climbing techniques in the Arcata community forest.

And generations of residents of Arcata learned skills of stream restoration by working on the small streams that run through Arcata. A city that remembers its streams is a city that is attempting to heal itself. High school students, working with their biology teacher, are healing Jolly Giant creek, attempting to bring back salmon and steelhead runs by planting willows along stream banks and monitoring water quality. If their efforts can succeed over the next century in this small coastal watershed, perhaps there is hope for other watersheds. Watershed associations, dedicated to restoration of the integrity of habitat for salmon and steelhead, could begin healing this human-damaged landscape.

The City of Arcata is also officially opposed to exotic, invasive alien species. It is unlikely that exotic, invasive alien species can be eradicated from Arcata, but they can perhaps be kept under control. The Environmental Services Department of the City of Arcata circulated a poster "Wanted Dead" with color illustrations of four of the most noxious weeds found in the city including German ivy (now known as Cape ivy [*Delairea odorata*]), English holly (*Llex aquifolium*), Scotch broom (*Cytisus scoparius*), and Pampas grass (*Cortaderia selloana*).

Over the past four decades, Arcata has restricted the sprawl that blights so many other California cities. However, the ecological footprint of Arcata, as with any other city, extends far beyond the official boundaries of the city. Water for the City of Arcata is drawn by wells drilled in the Mad River a few miles upstream from Arcata. The Mad River rises in the mountains of the Yolla Bolly and flows through forests and ranchlands before entering the Pacific a few miles downstream from Arcata. There is a dam on the Mad River at Ruth, where the river emerges from the Yolla Bolly mountains. The water in the reservoir behind the dam is released during dry summer months to provide a steady flow of water for Eureka, Arcata, and for industrial use at the pulp mill on Humboldt Bay.

Natural gas, piped to homes in Arcata, comes from Alberta, Canada, through an elaborate pipeline transmission system. Electricity is generated in many places outside of Humboldt County and distributed through elaborate systems. Furthermore, Arcata residents and most residents of the Klamath-Siskiyou region, remain car dependent, and the fossil fuel economy brings food and merchandise into Arcata from distant regions.

During the early 1970s, some students on the HSU campus established the Campus Center for Appropriate Technology (CCAT). Fueled with enthusiasm for "appropriate technology," the students won permission from the administration to takeover an old house on campus and convert it to what then was a model energy efficient house with thermal curtains, solar panels, a compost toilet, rainwater collection system, and bicycle-driven washing machine. The students planted herb gardens and vegetable gardens around the house and opened the house for discussion groups on "organic agriculture" and "solar installations."

For a time, during the energy crisis of the mid-1970s, it seemed that a revolution was occurring. State and federal legislation mandated higher energy efficiency standards for new buildings. Recycling and waste reduction, once thought of as "radical" demands by environmentalists, are now mandated by state law. Conservationists in Arcata pioneered in the development of recycling. The Arcata Community Recycling Center, started in the back of the Northcoast Environmental Center in 1971, became a model for community-supported recycling. But the City of Arcata did not formulate and implement a curbside, citywide recycling system until the late 1990s.

However, truly radical changes in lifestyles in Arcata have not taken place. In the late 1990s in Arcata, advocates for "bicycle friendly streets" and an automobile free Plaza were rejected by the progressive, "green" city council. There are no serious attempts to radically reduce automobile use in Arcata.

When students, during the 1970s, used the slogan "small is beautiful," they failed to appreciate that "appropriate technology" must be nested in a framework of values. The central chapter in E. F. Schumacher's book, *Small Is Beautiful*, is entitled "Buddhist Economics." Schumacher frames questions of technology in the context of Buddhist precepts including "do no harm" and "right livelihood." Questions concerning technology are frequently phrased in terms of efficiency and effectiveness of the technology in question to achieve stated goals. When the narrow criteria of efficiency is used, however, without attention to these and other questions, then the decision concerning use of a specific technology becomes value-neutral.

Instead of value-neutral technology, E. F. Schumacher, Langdon Winner, and other critics of technology conclude that we must ask questions concerning the meaning of technology. Is the technology understandable by laypersons or is it understandable only by experts? Is the technology mutable, that is, can its form be easily changed when emerging circumstances require new approaches? What is the net energy output from the technology? When these types of questions are integrated into discussions of "appropriate technology" then a community is able to make an informed, value-laden collective decision concerning the incorporation of a specific technology in that community. The core issues involve our sense of place not the technocratic ideas of scientists. When issues are framed in terms of technocratic hypotheses, then the philosophical questions are ignored.

Certainly, there have been many attempts, over the last three decades, to make Arcata a more desirable place to live, a humane place to dwell. The result has been higher costs for houses in Arcata than other cities in the bioregion.

Residents of Humboldt County shop at national brand name stores at the Bayshore Mall shopping center in Eureka. Residents of Del Norte County shop at the local Wal-Mart and residents of Trinity County drive across the mountains to shopping centers in Shasta County, in the city of Redding.

The dilemmas that Arcata faces as it enters the twenty-first century reflects the cultural wars that swept across America from the 1960s through the 1990s. The "limits to growth" rhetoric of the 1970s was replaced with "sustainable development" rhetoric in the 1980s and 1990s. "Sustainable economic growth" is based on the mentality that favors increased human population, increased economic growth, and increased income as prerequisites to maintaining environmental quality.

In the "limits to growth" paradigm, smaller human population, steady state economics, locally grown food and locally made manufactured products are favored over any policies that would likely increase human population or human impacts on the bioregion.[2]

Will Arcata maintain its "limits to growth" paradigm over the next generation or succumb to "sustainable development?" In the first decade of the twenty-first century, the freeway from Eureka to McKinleyville, passing through Arcata, is glutted with traffic during rush hours as more and more people commute from the burgeoning bedroom subdivisions in McKinleyville to jobs in Eureka. Unless growth in McKinleyville is addressed, downtown Arcata will be merely a "historic district" and the real action for "economic growth" will be in McKinleyville. McKinleyville is to Arcata and Eureka what Santa Rosa is to Marin County and San Francisco—a boomtown spoiling all the best laid plans for "sustainability" in the region.

Part of the dilemma faced by Arcata is that real change is collective change, not just change based on individual lifestyles. Yet one of the main constituencies for "green" politics in Arcata consists of college students and ex-college students who chose to remain in Arcata. Many of them distrust government and distrust collective action dictated by government as much as any right-wing conservative distrusts government. Many college students, and ex-students, who gladly experiment with building their own straw bale houses, compost toilets, and other manifestations of "appropriate technology" also rail against government agencies who attempt to institute new policies encouraging such changes throughout the bioregion.

Although not inside the city limits of Arcata, the Arcata bottoms, lowlands from the Mad River to Humboldt Bay, are in the "sphere of influence" of Arcata. At the turn of the century dikes were built on the northern end of Humboldt Bay, thereby allowing the bottoms to be developed as rich dairy farms. Each farmer built a huge barn out of old-growth redwood. At the end of the twentieth century virtually all of these barns are obsolete, and many are abandoned in favor of new, metal barns. Yet these barns symbolize an agricultural legacy of a time when this bioregion was more self-sufficient in food production. The barns themselves are architectural treasures. Never will such large structures be built again with old-growth redwood. Indeed the lumber in the barns is still considered so valuable that some people propose "mining" the barns—tearing them down and selling the old-growth redwood in urban areas of California for redwood decks and paneling. Others suggest that either the barns be maintained or left to decay on site, like old-growth redwood trees that fall in the forest, these barns as they decay will nourish the soil of the Arcata bottoms for a new forest that might emerge over the next hundred years. The fate of these barns is symbolic of the fate of the bioregion. Will we continue to export the heritage of the bioregion or live within limits in this place?

Citizens of Arcata cannot change the politics of the bioregion by themselves. Arcata remains a kind of deviant community in Humboldt County. However, few other cities of 16,000 residents in California have received so much publicity for their "green" policies. The evolving relationship between residents and landscape, dwelling between Humboldt Bay and the Mad River, between the ocean and the mountains, can be creative and innovative. "Harmony with nature" and "balance between the environment and economy" were phrases popular in the 1970s. At the end of the 1990s both these phrases rang hollow. Short-term economic growth has not been balanced with ecological sustainability. Ecological sustainability must take precedence in all decisions if there is to be any economic sustainability. We don't live in "harmony with nature" or "in balance with nature." Nature bats last. We live within the restraints of nature. We dwell in an evolving landscape shaped by dynamic forces of wind, fire, water, and climate change. Will Arcatans, as a community, adapt to these dynamic forces of nature while protecting the integrity of wetlands, watersheds, and forests? Will Arcata continue to improve the quality of life of all its citizens while reducing its ecological footprint? Are residents of the region willing to take the drastic actions necessary to reduce their consumption to the minimum amount of land and resources to sustainably support a small and declining human population? These are questions that citizens must address in their vision for the twenty-first century.

Bioregional autonomy means that crucial political, economic, and social policy decisions are made at the local level. Bioregional autonomy could be based on principles of consensus decision-making, democratic decision-making, decisions made by local warlords or other types of decision-making. Devolution of power from international, national, and state government to local agencies, including county and city governments in the Klamath-Siskiyou bioregion, would come with costs. Because, over the last hundred years, the timber, fisheries, and the soil of the region have experienced drawdown, that is, have been exploited and shipped to national and international markets, the richness of Nature has been reduced. Human labor, money, and creative efforts are needed to enhance the possibilities of material subsistence in the Klamath-Siskiyou bioregion.

Technology continues to be controlled by national and international organizations. Bioregional autonomy requires local knowledge, specific to place, that is used within the bioregion based on principles of equality and sharing of knowledge.

NOTES

1. Dave Foreman, co-founder of Earth First!, promoted an ecocentric viewpoint in his writings and actions in relation to Nature. Widely known for his support of the protection of the North American wilderness and biodiversity, Foreman (1991, 1992, 2004, 2011) coordinated the Wild Lands Project. He also supported direct action to conserve and protect the ancient forests of North America, especially the forests of the Pacific Northwest. Ecotage was the order of the day with environmental activists performing acts of tree spiking, tree sitting, and the blocking of logging roads to prevent the timber trucks from ferrying the cut logs to the timber mills or for direct export to Asia.

2. The Limits to Growth paradigm was widely debated and deliberated worldwide following the series of reports commissioned by the Club of Rome (founded in April 1968 comprising of eminent persons from all over the world), with the first report being entitled *Limits to Growth* alerting to the limits to the world's resources. The second report titled *Mankind at the Turning Point* announced that the limits are within the control of humans. Later updates continue to sound the alarm bells as well. See Mesarovic et al. 1974; Meadows et al. 1972, 1992, 2004.

Chapter 5

A Geography of Hope

Restoration in Redwood National Park

Lyon Ranch is a historic sheep station on the slopes of Schoolhouse Peak, the highest mountain in lower Redwood Creek.[1] At the end of the nineteenth century, several thousand sheep roamed through the meadows and oak savanna where Chilula Indians hunted elk and deer before 1850. The wool from Lyon Ranch was transported to Eureka where it was shipped to San Francisco.

Sheep ranching ended during the 1970s. It was no longer economically feasible to raise sheep for wool in this region. The ranchlands were acquired by the federal government and added to the Redwood National Park. Two decades later, native wildflowers are beginning to emerge again including blue dicks (*Dichelostemma capitatum*), inside out flower (*Vancouveria planipetala*), and miniature lupine (*Lupinus bicolor*).

Oak savanna on the Lyon Ranch is now recognized as one of the most significant and important remaining examples of this type of oak savanna landscape in the northcoast basins region.

When fire returns to the landscape, wildness returns. During late August 1998, I observed the Park Service burn 251 acres of Schoolhouse Peak. This was the first prescribed burn of the season. The Park Service set a goal of burning 2,700 acres during this brief burning season. Stephen Underwood, vegetation management officer for Redwood National Park, said that his crew only gets fourteen to twenty days a year when the humidity, wind, burn conditions, availability of fire crews from California Department of Forestry all come together for a permissible, prescribed burn day. On this day, the humidity ranged from 50 percent to over 60 percent, and the temperature was in the mid-seventies, light wind, no cloud cover.

The objectives for the burn I observed were guided by the Fire Management Plan of Redwood National Park. The overall objective of the program is to reintroduce fire as a natural process in the grasslands of Redwood valley. The specific objective was to kill small, less than 6-feet high, Douglas fir trees so they do not encroach on the meadows and oak savanna, converting it into a Douglas fir dominated forest.

There are over 6,000 acres of oak savanna in the Redwood National Park. Studies indicate these oaks tend to regenerate from root sprouts rather than from acorns, thus sending a light burn through the oaks does not damage their reproduction.

Park scientists have concluded that if the meadows are burned in the fall, it helps native grasses to reproduce, and after burning, greater biodiversity of native plants has been noted. Burning also helps kill scotch broom, an exotic invasive species proliferating in the meadows along the Bald Hills.

Late in the afternoon, I drove to the fire tower atop Schoolhouse Peak for another perspective on the prescribed burn. Surrounded by the fire blackened mountain, with smoke still blowing across the fire tower, I realized how our perceptions of forest fires have changed over the past decade. Smokey the Bear's slogan, "only you can prevent forest fires," has changed to acceptance of fire as a natural process in the landscape. Used wisely, prescribed burning could bring back to the landscape something approximating a natural fire rhythm. However, if human managers are necessary during the next century in this watershed to restore wildness during that time frame, then can a restored "wild" landscape then escape from management by humans?

Bringing back the abundance of this watershed will require patience, intelligence, and time. Healing is an art, as much as a science, a commitment as much as a task. It requires sensitive attention to the needs of the lands.

Very little attention was given to wildness and restoration on a beautiful autumn day in October 1969, when a past, a present, and a future president of the United States converged on the small community of Orick, near the mouth of Redwood Creek, to dedicate the Redwood National Park. Lyndon Baines Johnson, Ronald Reagan, and Richard Nixon, along with Lady Bird Johnson, for whom the dedication grove was named, added their praises to those who had preceded them in admiring the aesthetic value of ancient redwood groves. But they came to celebrate a political compromise, not to celebrate ecological sustainability.

The most intense parts of the battle to establish a Redwood National Park occurred between 1964 and 1968. When President Lyndon Baines Johnson signed the Redwood Park Act, the major environmental groups, including the Sierra Club, thought the battle for the redwoods, a battle that stretched across the twentieth century, was over.

On the October morning I drove from Arcata to Orick along the Redwood Highway to observe the prescribed burn in Redwood valley, I stopped at the Palms Cafe for breakfast and contemplated the absurdity of planting palm trees among the ancient redwoods. Across the street from the Palms Cafe was the Lumberjack bar, home of ardent loggers and in 1969 owned by Pat Dorsey, who was an eloquent spokesman for the loggers intent on clear-cutting the last old-growth redwoods in Redwood Creek.

Proposals for a Redwood National Park had been floated since the 1930s. Indeed, during the Great Depression, the federal government began buying redwood forest lands near the Klamath river from bankrupt timber companies. What was called the Northern Redwood Purchase Unit contained thousands of acres of old-growth redwood forest by 1940 when the program was ended due to the emergency of World War II.

The national campaign to create a Redwood National Park began in 1963 with the publication by the national Sierra Club of an exhibit format book, *The Last Redwoods: Photographs and Story of a Vanishing Scenic Resource*, and in 1964 with the publication of a report prepared by the National Park Service, *The Redwoods: A National Opportunity for Conservation and Alternatives for Action*. In that document, the National Park Service acknowledged that roadbuilding, logging, agriculture, and so-called conversion of forests to pasture were contributing to the destruction of old-growth redwoods, and they proposed four alternative plans for a Redwood National Park.

Lucille Vinyard, who helped establish a local group, Citizens for Redwood National Park, remembers those days: "It was an exciting time in 1964. We went on field trips with timber companies, the National Park Service, and politicians. We were engaged in discussions over the most appropriate location for a Redwood National Park. We had constructive conversations. We quickly rejected Humboldt Redwoods State Park and the Klamath River alternatives and the real conflict was over whether the park should focus on Mill Creek in Del Norte county or Redwood Creek in Humboldt county."

During the early 1960s conservationists settled on Redwood Creek as the site for the proposed Redwood National Park after the National Geographic Society announced the discovery of the tallest redwood tree in the world in Redwood Creek in 1964. But the biggest battle of the redwoods during the early 1960s was the proposed routing of US 101 freeway along famed Gold Bluff Beach, north of the hamlet of Orick. The 1960s were boom times for freeway building in California, and the powerful California Department of Transportation proposed either cutting a swath through the old-growth forest in Prairie Creek State Park or routing their proposed freeway on the beach near Fern Canyon, a favorite tourist attraction featuring walls hung with five-finger fern.

Conservationists proposed a third alternative, on the east side of Prairie Creek State Park. Recognizing that the eastside route would be extremely expensive, conservationists assumed that such a freeway would never be built because it would traverse some of the most erosive terrain in the region.

However, this proposed routing came back to haunt conservationists in 1978 when the Redwood Park Expansion Act was passed by Congress. That act included a provision authorizing construction of the eastside freeway. The eastside freeway was opened a decade later. It cost over $300 million to build seven miles of little used freeway crossing a 1,600-foot ridge. During construction of the freeway, massive slides from improper construction dumped silt into salmon-spawning streams. But the freeway boondoggle was decades in the future during the bitter fights over establishing Redwood National Park that erupted between 1964 and 1968.

The irony of the political compromise that created the first Redwood National Park in 1968 was that more acres of old-growth redwoods were traded away to timber companies than were included within the park boundaries. The national Sierra Club proclaimed a victory with the passage of the 1968 Redwood National Park Act, but the logging companies walked away with the cash and the right to log even more redwood forests.

Some local conservationists knew that the compromise bill creating the park was not ecologically justified. A small group of HSU students held up signs at the dedication ceremony. The signs said "fatten the worm" referring to the narrow corridor up Redwood Creek eight miles to the Tall Trees Grove where the tallest known redwood in the world is located. The young protesters at the Redwood Park dedication were thinking like a watershed. At that time, no significant complete watershed had been preserved for baseline comparison with logged watersheds, and none could for all have been at least partly logged.

I hiked up the creek to the Tall Trees Grove in January 1970, fording Redwood Creek with water up to my waist. Upslope, loggers were clear cutting ancient redwood groves outside the narrow corridor included in the first park bill. Sounds of chainsaws and logging trucks filled the air, drowning out the sounds of the creek. I hiked with Rudi Becking, a forestry professor at HSU, and several of his students. Becking was studying the relationship between logging, roadbuilding, and siltation in Redwood Creek. With his inspiration, students formed the Emerald Creek Committee. This grassroots, local group led the struggle to expand the Redwood National Park from ridge to ridge on the lower eleven miles of Redwood Creek. Activists trespassed through the lands of Simpson, Arcata Redwood Company, and Georgia-Pacific to document the damage done by logging. They also encouraged state and federal agencies to conduct studies of the sediment load pouring into the floodplain of Redwood Creek from upslope logging.

The battle over expansion of Redwood National Park was a battle between those who think like a watershed and those who confused the forest for the trees. Proponents of park expansion realized that state and federal regulations would neither significantly stop the rapid destruction of old-growth redwoods nor protect landscape integrity and salmon habitat in Redwood Creek or the integrity of the landscape. Conservationists concluded that the federal government had to acquire the whole Redwood Creek watershed from Schoolhouse Peak to Orick.

The campaign to expand Redwood National Park was opposed by many local politicians who promoted the slogan "Don't park our jobs." The bitter political battle came to a focus in 1976 when Representative Philip Burton brought his House Parks Subcommittee to Eureka for a public hearing on the park expansion proposal. The mayor of Eureka, Sam Sacco, led the Mothers and Fathers of Eureka in a massive march from the county courthouse to the civic auditorium for the public hearing. The Sierra Club and other conservationists were denounced as "un-American." However, Representative Burton, a San Francisco Democrat who developed his devotion to preservation of ancient redwoods through his friendship with Sierra Club leader Edgar Wayburn, was determined that Congress would pass his bill—which was accomplished after fellow Democrat Jimmy Carter was elected president in 1976. Carter promised, during the presidential campaign, to promote the expansion of Redwood National Park.

When logging was stopped by the passage of the Redwood National Park Expansion Act, in 1978, the National Park Service was faced with the task of protecting the Tall Trees Grove and working with a severely damaged landscape. After approximately thirty years of intensive logging in lower Redwood Creek—from the early 1960s through 1978—some experts predict it will take hundreds of years for healing to occur in the watershed.

The upper Redwood Creek watershed remains mostly in private ownership, with the small exception of some parcels owned by BLM at Lacks Creek and a small part of the upper watershed within Six Rivers National Forest. The private lands of the upper watershed continue to be extensively logged and roaded and subdivided. Under provisions of the 1978 Redwood National Park Expansion Act, the National Park Service has limited responsibilities for approximately 30,000 acres upstream from the park boundaries, including the responsibility to comment on proposed Timber Harvest Plans submitted for approval to the California Department of Forestry (CDF) under provisions of the state forestry act. As with many other watersheds in the bioregions with many private landowners, there is always potential for conflict between the public interest in long-range, cumulative impact of human activities in the watershed and the short-term interests of landowners who want to build more roads, log timber, and hunt wildlife.

On another fieldtrip to Redwood Creek in late November 1991, this time with geologists working with the National Park Service, I walked along a "restored" slope of a small stream. During the early 1970s a timber company bulldozed a major road across these slopes in order to remove ancient redwoods. Fifteen years later, in place of ancient redwoods, there are thick stands of alder and young Douglas fir.

A survey by the National Park Service found over 300 miles of logging roads and at least a 1,000 miles of skid trails inside the expanded park boundaries.

On our fieldtrip, a supervisory geologist for the Park Service explained the process of roadbuilding and road removal. She said this is the largest project of road removal the Park Service has undertaken. Workers cleared a stream crossing and pulled out several hundred yards of road fill. The steep slope has been returned to its original contour. Logs found at the bottom of the road fill have been scattered on top of the recontoured slope. Downed trees and decaying logs are part of the structure of old-growth forests in the Pacific Northwest. Placing the logs on the restored hillside may help restore connectivity for various fungi and other plants from the surrounding forest onto the former road.

A D-9 bulldozer, a very large piece of diesel-burning equipment, sits on the road at the end of this project area. The raw earth is clay-like and clings to our shoes even though this is still the dry season.

The geologist explains that the darker-colored soil, the original topsoil, was pulled out of the stream crossing by the bulldozer. Native plants are likely to sprout from the topsoil, probably from seed still in the soil.

The Park Service has learned many lessons during the nearly two decades of experiments in restoration in the park. During the first few years, after park expansion, the Park Service contracted with small contractors, mostly ex-students at HSU who had been trained in appropriate technology. These students believed in the then popular "small is beautiful" ideology and were opposed to big machines. They used picks, shovel, hand saws, and wheelbarrows to build check dams along eroding gullies, but most of the structures washed out during the first major storm event.

The Park Service geologists decided that big bulldozers had made these roads, and big bulldozers were the appropriate tools to take out the roads. Using D-9 bulldozers, the Park Service can move more earth, dig down through road fills, and restore the original slope of the hillside.

Restoration workers say they are like people working in the emergency room of a hospital. In a hospital, if the patient is bleeding, the first task is to stop the bleeding. The priority for restoration workers in Redwood Creek watershed is to slow the rate of erosion into the stream bed by clearing potential erosion sites such as log decks and road stream crossings. The restoration workers always have choices and always have to establish priorities.

Actions undertaken during restoration may not always be aesthetically pleasing, and the long-term results will not be realized until long after the restoration worker is dead. Thus it is necessary to monitor restoration projects for many years, to carefully document the work so that later generations of restoration workers can reconstruct the sequence of change at a particular site and to explain why certain actions were taken at certain times.

A restoration worker is careful with the timing of actions. Modesty is a great virtue for restoration workers. Restoration workers are conscious of, and responsive to, natural events such as floods, earthquakes, landslides, and windstorms.

Restoration means recovering the geography of hope and thinking over a longer time frame than one's own lifetime. What makes a "good restoration worker"? "Ask why and ask why again," says our Park Service guide. "Why are we doing this action? Is our motivation true to the long-term function of this slope? Why take out this old road at this time? Why plant these trees at this time? Why not wait a while and see what happens before rushing in where fools have rushed before?" A restoration worker should be an observer and an experimenter, a generalist who studies wildlife, fisheries, geology, natural history, sociology, ecology, and ecophilosophy. Generalists understand scientific and philosophical theories but know that all concepts and constructs are nets. Many of the theories are not fine enough nets to catch all the circumstances of a particular place at a particular time. The practice of restoration work requires a kind of openness, not a bureaucratic play-by-the-rules approach. It requires creativity, watchful awareness of the impermanence of the watershed, and innovation. In this watershed, a restoration worker looks at the geomorphology, the sources of erosion, and potential erosion. Restore the soil. That is the slogan of restoration workers in Redwood Creek.

But can we deliberately create the conditions that restore wildness to lower Redwood valley? Wildness and wilderness as ideas, and as physical conditions on the land, have been criticized by some feminists, postmodern writers, and some American Indians. Chilula Indians dwelled in lower Redwood valley for hundreds, perhaps thousands, of years before contact with Europeans in the middle of the nineteenth century. There is some evidence, through oral histories and archaeological work, that the Chilula "managed" parts of lower Redwood valley not only by using prescribed burning to maintain the meadows as elk and deer habitat but by other techniques as well.

Can human management increase the possibility for wildness to return in this, or any other, watershed? Wildness, by definition, means undomesticated, unmanaged by human stewards, the free-flowing energy of a place. Industrial forestry management greatly diminishes wildness. A clear-cut is a wound on the landscape. A valley with over 300 miles of logging roads is a wounded landscape. But the possibility for wildness remains. If it is possible to restore the self-regulating processes of wind, fire, flood, and predator-prey relationships, then wildness can return to once human-dominated places.

Playing with Thoreau's dictum, "In wildness is the preservation of the world," we can list some of the attributes of wildness—solitude, ability of a place to maintain the flow of evolution, less human control rather than more human control of the landscape, and habitat for wild creatures. With the application of these attributes to lower Redwood valley, wildness is returning to the valley.

Officials of Redwood National Park have posted signs at trail entrances warning visitors they may encounter mountain lions and bears. When big predators return to a landscape once devastated by human activities, is wildness returning? Not necessarily. Bears and mountain lions sometimes wander through the suburbs of Arcata and Eureka. However, the presence of big predators in Redwood Creek is one indicator of wildness.

Hiking through Redwood Creek above Tall Trees Grove, to Rocky Gap, during late summer is like hiking to the center of the world, the inner sanctum, the gathering place. The flow of water in Redwood Creek in late summer is barely a trickle but at Rocky Gap there are deep pools of water among huge boulders. Swimming in the waters under a late summer sun, lying on the sandbars, sleeping under the stars, my companions and I wondered if mountain lions were watching us from the forest. Tracks downstream from our camp which we found on the river bank in the morning testify to the presence of these creatures. I have never seen a mountain lion in my many trips to Redwood Creek, but I have felt their presence.

NOTE

1. For a detailed study of the ancient forests of the Pacific Northwest, see Elliott Norse (1990). For other histories on the logging of the redwoods and the battles to save them, see Delong (1968) and Schrepfer (1983).

Chapter 6

Grassroots Restoration and the Culture of Reinhabitation

Mattole River

I began my pilgrimage to the Mattole watershed at the Redwood Monastery near the headwaters of the Mattole River. From Garberville I took the county road west, toward Shelter Cove and turned south at Whitethorn onto another county road. Whitethorn was like a sanctuary during the late 1960s and early 1970s. Hippies, pot-growers, back-to-the-land people, urban dropouts, and leftover loggers found refuge here. The Redwood Monastery, however, set in a grove of redwoods is a different sort of sanctuary—a spiritual anchor in a sea of nihilism. The nuns practice a strict rule of silence for twelve hours each day, but after supper they talked avidly about logging practices, siltation of salmon-spawning gravel due to roadbuilding, and the spiritual reasons for protecting old-growth redwood forests.

Salmon and steelhead have sought the cool running waters and clean gravel in the upper Mattole for millennia, and this sanctuary, this refugia, was under attack from loggers interested in values other than the intrinsic value of ancient stocks of salmon. The nuns wrote letters to legislators, talked with visitors to the monastery, and asked me to visit with them because they were seeking advice on how to proceed with their activist meditation for the redwoods and salmon. We talked until vespers, and although we didn't resolve the questions that filled our minds, I went to bed filled with inspiration and hope.

Salmon returning to their ancestral waters after spending years at sea find many obstacles from the mouth of the Mattole to the spawning gravels where they instinctually want to spawn and then die. The decline of the salmon is a sad story of wasted opportunities, denial, and lack of effective government regulation of roadbuilding, logging, grazing, and other human activities that degrade the vitality of a watershed. The degradation of the Mattole watershed also illustrates the cumulative impact of decisions made by landowners over several decades. Each landowner, acting in his or her self-interest, cumulatively logged an estimated 97 percent of the mature forests of the watershed between 1947 and 1987. Freeman House, a longtime resident of the Mattole watershed, in his book *Totem Salmon: Life Lessons from Another Species*, documents how sustainable salmon fishing based on traditional American Indian rituals were overwhelmed by government-sanctioned fish harvesting in the Mattole.

Faced with continuing degradation of the remaining salmon-spawning areas of the Mattole, a local group called Sanctuary Forest convinced two logging companies—Eel River and Barnum Timber—to sell 200 acres of mostly virgin timberland to the Sanctuary Forest Trust. Using a combination of public and private funds, the trust was able to protect nearly 3,700 acres of forest through purchase and almost another 6,000 acres through conservation easements.

The road down the Mattole is a dirt and gravel road following the ridge, through the old community of Ettersburg. The Mattole valley is defined on the east by Gilham Buttes and on the west by the King Range.

Gilham Buttes is a local landmark in the upper valley. Some local residents are attempting to create a local land trust to accept conservation easements from local landowners to provide a wildlife corridor between Humboldt Redwoods State Park and the Kings Range National Conservation Area (BLM). Based on a model developed by a national organization, The Wildlands Project, a group based in Arcata called Legacy: The Landscape Connection is developing detailed maps for all the northcoast basins utilizing Geographic Information System (GIS) and on the ground surveys of wildlife to establish a strategy for protection of native biodiversity based on principles of conservation biology.[1]

The upper Mattole road drops off the ridge in a series of switchback turns and reaches the river near its confluence with Honeydew Creek. Honeydew is the largest creek on the eastside of the King Range. The upper reaches of Honeydew contained some of the last old-growth forests in the Mattole watershed until it burned in a massive wildfire.

Wide gravel bars and steeply eroded slopes define the Mattole from its confluence with Honeydew Creek to the mouth of the Mattole near the community of Petrolia. At the community of Honeydew, an old country store sits on the banks of the river where the Panther Gap road converges with the Mattole road. Local residents hang out in front of the store, sharing gossip. During the summer, Sunday afternoon softball games are played across the road from the country store.

During the 1930s the Mattole was known by fishermen for its steelhead and salmon runs and for its apple orchards. There were small logging operations, but no massive clear-cutting of Douglas fir. During the 1970s and

1980s urban refugees bought 40 and 80 acre parcels of logged-over land from ranchers and timber companies. Some of these "new settlers," as they are still called by "oldtimers" in the valley, have matured from a naive back-to-the-land mentality to a bioregional commitment. They have taken salmon as their totem animal and have dedicated themselves to salmon restoration and community building. Their ground-up approach to restoration, they claim, is more adaptive to the needs of the watershed then the top-down regulatory structure of the State of California. These "new settlers" looked at the old culture of the Mattole—the rural culture of sheep and cattle ranchers and remnant loggers who continued to live in the valley, the state of the watershed (publishing an analysis that defined the sources of siltation in the creek), and the state of the salmon and steelhead populations. And they went to work, building community, talking with ranchers and other landowners about their needs, seeking grants from the State of California for salmon restoration, and creating a new culture. Some of them wrote a play, "Queen Salmon," and toured the state with the play—a story of salmon bringing together loggers, new settlers, and ranchers as they all come to realize that salmon is their totem animal. They looked at the river through the eyes of salmon and steelhead.[2]

They formed the Mattole Restoration Council whose mission is "the restoration of natural systems in the Mattole River watershed and their maintenance at sustainable levels of health and productivity, especially in regards to forests, fisheries, soils and other plant and animal communities."

The mouth of the Mattole is one of the most remote places on the California coast. Just to the north is Cape Mendocino. Just off Cape Mendocino is the famous Triple Junction. South of the Mattole is the Lost Coast, a rugged unroaded coastline on the western slopes of the King Range.

Salmon, seeking their ancestral home, must discern how to find this obscure small river mouth along the vast coastline of the continent. Entering the river mouth, into the lagoon, they rest before their journey upriver to spawn and then die. But their journey has become increasingly perilous because of human-induced changes in the landscape.

The new settlers devoted their attention to the watershed, attention to the salmon, attention to the whole system that they called home. They used rotary screw traps (a device to trap and release fish so they can be counted) and systematic surveys of the state of the river system. They began to develop a sense of their mission and a list of priority projects for restoration.

In the offices of the Mattole Restoration Council began organized restoration efforts in the Mattole watershed in the early 1980s. Some of these efforts were made by landowners, others by the Mattole Restoration Council and other groups. In 1989 the council published *Elements of Recovery* which was a comprehensive survey of human-caused disturbances and natural stresses in the Mattole watershed.

During the first decade of the twenty-first century, Nature is taking care of some of the problems herself. Human efforts can enhance the recovery or healing process. Healing will never be complete—that is, the watershed will never return to a previous state. However the protection of remaining refugia such as remaining areas of excellent spawning gravel for salmon and remaining stands of old-growth forest, as well as application of energy, time, and money—always limited resources—to high priority projects can enhance the possibilities of recovery in the system.

In their process of identifying priorities for restoration projects, the Mattole Restoration Council looks for intact habitat areas or areas that have one or two threats. For example they look at tributary streams that are low in sediment, an indication of the use of tributaries by salmonids—salmon and steelhead—and tributaries that have potential for more rapid recovery. We must accept the fact that it will take twenty to two hundred years before stabilization occurs in some regions of the Mattole watershed. A certain level of disturbance is natural. The rate of disturbances was changed by humans during the twentieth century in this watershed with roadbuilding, clear-cutting, grazing, extermination of native predators, and introduction of exotic alien species such as exotic grasses.

The Federal Bureau of Land Management (BLM) controls portions of the Mattole River watershed in the King Range. Most of the BLM managed land is designated as "wilderness" under federal legislation. Major forest fires on federal lands combined with the removal of some roads on federal lands have altered the upper Mattole watershed during the last two decades.

All efforts at restoration will come to naught unless the attitudes of landowners, regulatory agencies, and tourists change to favor the objectives of long-term restoration rather than exploiting whatever they can at the present time from the remaining resources of this watershed.

The council has contracted with state and federal agencies to do numerous projects in the watershed including a road-removal project in the King Range National Conservation Area, tree planting projects, stream bank stabilization projects, and fish monitoring projects.

The Mill Creek Conservancy was organized to work for the acquisition of the Mill Creek watershed from a logging company that wanted to log most of the timber from that watershed. In a complex deal, most of the land in the watershed was acquired from the logging company and transferred to the BLM to be managed under the general management goals and objectives of the King Range National Conservation Area. The county roads department installed a new culvert on the county road crossing Mill Creek, but the way the culvert was installed prevented salmonoids from crossing through the culvert to reach spawning gravel upstream. The Mattole Restoration Council built a series of graduated pools and rockwork so that salmonoid could jump into one pool and then into a higher pool and find their way through the culvert.

Restoration workers learn to think like a salmon, and their efforts are intended to benefit the ability of wild salmon to reinhabit their historic spawning grounds.

Reinhabitation is a major theme of the bioregional movement as expounded in the Mattole watershed by such leading resident thinkers as Freeman House and Peter Berg. Reinhabitation has many facets and involves work over generations. Therefore schooling for the next generation is an important aspect of the work of those attempting to create a bioregional culture. The Petrolia School, based on property near Mill Creek watershed, provides a semester-long curriculum for mature high school students on ecology, restoration, and sustainable rural living. Students are given opportunities for hands-on learning by working as apprentices on restoration projects.[3]

Reinhabitation means getting to know and respect and respond to changes in the homeland. In the words of the "Ecoforesters Way," a statement developed by foresters, bioregionalists, and philosophers committed to sustainable forests, the question is not how much we can take, but how much we can leave. Corporate owners, on the other hand, usually ask how much we can take in order to sustain short-term profits for the corporation.

Whatever the outcome of reinhabitation and restoration, the work itself is what poet/philosopher Gary Snyder calls the "real work." This is the work of healing, helping, hands-on experience, dwelling in the watershed. Science is important, but science is not the whole project. Reinhabitation requires a spiritual commitment to the life force of a place, a commitment to work in community with other humans, and with salmon and other native species, in ritual affirmation of the integrity of life and death.

NOTES

1. For an intensive study of the biodiversity crisis occurring in the Pacific Northwest and on ecosystem management utilizing conservation biology principles, see Grumbine (1992).

2. Freeman House (1999) has documented the restoration activities of the Mattole watershed and the rehabilitation of the salmon.

3. For further discussion of the concept of reinhabitation, see Berg (1978, 2009) and Berg and Dasmann (1978).

Chapter 7

The Last Battle over Old-Growth Redwoods

The Headwaters Forest

We walked hand in hand along Fisher Road, near the old logging town of Carlotta, toward the gate that marked the property line for Maxxam (PALCO).[1] Joanna Macy, deep ecology teacher and author, was my companion. Behind us, a Buddhist monk chanted a traditional Tibetan chant. The line of citizens stretched for miles along this country road. Clouds hung low over the hills on this September morning, silent companions to the forest primeval. We came to speak for the integrity of the forest, demonstrating with our feet, drumming out a message to Charles Hurwitz that we want to save Headwaters. Hurwitz, a Houston (Texas)-based financier, purchased controlling interest in PALCO in 1985 using high interest bonds called "junk bonds." Hurwitz made his position very clear the first time he visited Humboldt County after he bought control of PALCO. He told the assembled employees of PALCO, his employees now, "I believe in the golden rule. He who has the gold rules."[2]

Hurwitz may rule with money, but thousands of citizens are demanding something more important than money, the rule of law—the laws of ecology and the laws of the United States. Citizens are sending their message to Hurwitz, via the media, and to the government agencies that are supposed to protect the spotted owl, marbled murrelet, and coho salmon by enforcing provisions of the federal ESA. Humans have no right to destroy the habitat of other creatures, especially of endangered species. There are other values besides the value of the almighty dollar. These lands are our lands, people of the land, people of the bioregion. Protesters are saying, "You, Charles Hurwitz, have no right to destroy the integrity of our bioregion." Some protesters carry signs inscribed with the slogan "Save Headwaters, Jail Hurwitz."

As the long silent line of protesters reached the locked gate, they faced a solid line of police, drawn from the Humboldt County sheriff's department and many other police agencies in the region, who are determined to arrest protesters for trespass if they crossed the line. Each person made a decision—cross the line drawn in the mud and be arrested by sheriff's deputies, or stand aside and witness in other ways for the integrity of the ancient redwood forests.

Many made the decision to cross the line in order to witness for higher laws—the laws of ecology and the intrinsic worth of nature. Some who crossed the line had been arrested many times before, arrested for demanding environmental justice, arrested for demanding that Charles Hurwitz be brought to justice for his crimes against this place, his crimes against the integrity of the ancient redwood forests. For others, this was the first time they have been arrested for trespass—"forgive us our trespasses Lord, as we forgive those who trespass against the sacred forest." Crossing the line in twos, fours, dozens, then by the hundreds—law-abiding citizens—doctors, teachers, workmen, students, retired people—were arrested, handcuffed, thrown into steel cages, cited for trespass by sheriff's deputies, and deported from the forest. More and more people crossed the line to witness for the forest. They crossed the line in nonviolent protest, in the spirit of Gandhi, in the spirit of wilderness, in the spirit of the forest, crossing over with tears in their eyes pleading with the police to spare the ancient redwoods, pleading with police to defend these great sentient beings, pleading with Charles Hurwitz to save these ancient forests.

Almost one thousand people were arrested by the time darkness fell over the ancient redwoods during that long September day in 1996—the largest number of arrests ever at demonstration for the forest in the history of the United States. In the gathering darkness even the police were overwhelmed, exhausted by the strength of the protest. They took arrestees in police buses and dumped them on remote stretches of the public highway east of the protest site.

I made trip after trip in my jeep, picking up stranded citizens who had been arrested and released. They were tired and thirsty, but exalted by the presence of the great energy of the ancient forest. Into the night as more and more citizens are released from captivity, they collected at the Earth First! base camp near the Van Duzen River, collected around the blazing campfire, collected around the drummers, drumming out the message "save the forest, save the forest." In the cool September night, fog moved into the forest, comforting the ancient trees, nourishing them in the late summer drought, just as protesters were protecting them with tears of compassion.

It was the fifteenth of September, end of the nesting season for marbled murrelets. The marbled murrelet is an endangered bird that nests in the ancient redwoods of the Headwaters Forest. Under provisions of the U.S. Endangered Species Act, loggers were ordered by the courts not to enter this forest until the end of the marbled murrelet nesting season. Now, the loggers are revving up their chainsaws. Empty logging trucks begin moving from PALCO's Scotia mill toward the Headwaters Forest, loggers invading the sacred groves where the golden rule is the only rule—those with the gold rule, dominating, clear-cutting, attacking the last stands of these Monarchs in the Mist. And the forest weeps in sorrow, weeps for the pathetic actions of cruel humans who don't realize that self-realization means realizing our greatness in modest participation in the life of the forest primeval.

Hurwitz bought PALCO in 1985 for $870 million. For decades previous to the Hurwitz takeover, PALCO had been considered the most conservative logging company in the redwood ecosystem. Opened in 1887, holding almost 190,000 acres of multi-aged redwood forests, PALCO was considered a local company rather than a mega-corporation such as Simpson, Georgia Pacific, and Louisiana Pacific who owned most of the industrial forestlands of the northcoast basins of California. The "old" PALCO logged old-growth redwoods, but at a rate much slower than any other logging company in the redwood region.

Immediately after purchasing the company, Charles Hurwitz increased the rate of cutting on PALCO lands by 300 percent. He sold many of the assets of the company, liquidated the workers' pension plan, and aggressively attacked the integrity of the forest system.

When a photographer and forest activist, Greg King, discovered the largest unentered stand of ancient redwoods on PALCO property in 1987, he knew he had to publicize this place. Through photography, dramatic sit-ins on platforms in ancient redwood trees, blockades of logging trucks carrying ancient trees to the mill at Scotia, and by witnessing for the forests in the halls of Congress and before the California Board of Forestry, grassroots activists witnessed for the plight of the forests. After years of protests and attempts to gain reform from conservative governors in Sacramento, forest activists took to the streets and to the ballot box in 1990. A group of activists, mostly from Humboldt and Mendocinio counties, succeeded in getting a forest policy reform initiative on the statewide California ballot of November 1990. Called Forests Forever, this initiative provided $200 million in state bonds to acquire old-growth redwood forests from PALCO, reformed the state forest practices act to make policies more attuned to principles of ecosystem management, and opened the state Board of Forestry to more balanced representation from various stakeholders.

At the same time, April 1990, Earth First! activists Judi Bari and Darryl Cherney were organizing a massive summer of demonstrations in the redwoods. Modeled after Mississippi Summer, a campaign during the civil rights movement when thousands of northern college students went south to show solidarity with the frontline workers in the civil rights movement, Redwood Summer was to be a showpiece of nonviolent direct action.[3]

However, the summer began with an act of violence against the nonviolent activists. A pipe bomb was placed in the vehicle carrying Judi Bari and Darryl Cherney in Oakland on a spring morning in May 1990, as they began a tour of California college campuses to bring the message of Redwood Summer to urban college students. The bomb fragments tore into Judi Bari's flesh leaving her in pain and unable to walk without support for the rest of her life. Ironically, immediately after the bomb attack, the Oakland police arrested both Bari and Cherney on the charge of carrying a bomb. The police arrested them in the hospital where they had been taken for treatment. Many commentators feel that the FBI and the Oakland police were attempting to discredit Redwood Summer before it really got started by arresting two of the most credible leaders. The charges against Bari and Cherney were ludicrous and were dropped when they reached the district attorney's office, but "who bombed Judi Bari?" remains an unanswered question.

Even with Judi Bari in the hospital, Redwood Summer continued to unfold with a massive demonstration at the Georgia Pacific mill in Fort Bragg, repeated demonstrations against Louisiana-Pacific operations in Humboldt County, and concluding with a march on Labor Day weekend down the main street in Fortuna.

In the woods, actions including tree sits in PALCO's Headwaters Forest brought national media attention to the plight of the ancient forests.[4] Protesters demanded that the forests be given standing in human communities. As the Lorax said, in Dr. Seuss's famous classic, "I speak for the forests, I speak for the trees." Some loggers in Laytonville, however, demanded that the *Lorax* be banned from the public school library because they claimed it published radical environmentalist propaganda.

In the November election of 1990, citizens put an initiative on the ballot called the Forest Forever initiative. The initiative would create new rules for forest management on private lands in California based on "sustainable forestry" principles. The initiative lost by a narrow margin only after logging corporations put their own initiative on the ballot and after they spent over $8 million in opposing the Forest Forever initiative.

During the 1960s, during the battle over the first Redwood National Park Act, then Governor Ronald Reagan is reported to have said "When you've seen one redwood tree, you've seen them all." Cynical attitudes among many elected officials remained a theme of the politics of redwoods during the 1990s. In November 1997, Congressman Frank Riggs called citizens who were engaged in nonviolent protest demanding protection for ancient redwood ecosystems, "terrorists" and applauded police for dabbing pepper spray on the eyes of four teenage women who had chained themselves in nonviolent protest to a symbolic redwood stump in Riggs's congressional office in Eureka.

The battle over Headwaters Forest, from one perspective, was a battle over law and order. From another perspective, voiced by the supporters of the "wise use" movement, the battle was over private property rights and protection of private property of logging corporation from trespass by "radical" environmentalists.

For conservationists, the battle was over protection of public resources. Under ancient legal conventions, wildlife is under the stewardship of the state. Plants and animals have intrinsic value and value to the community of humans both present and future.

The Garberville based group, EPIC, defended the public trust doctrine and demanded that applicable state and federal laws be enforced in the forests and watersheds of the northcoast basins region of California since the late 1970s. Landmark court cases, including *EPIC vs. Johnson*, established that cumulative impact be considered in approving timber harvest plans (THPs) and in interpreting the ESA (Owl Creek case). In other words, citizen groups had to go to court time and again in a continuing campaign to get state and federal agencies to enforce laws that those agencies are mandated to enforce.

To see the bleeding wounds on the landscape caused by actions of PALCO, one only has to rent a small plane and pilot at Murray Field in Eureka, fly up Freshwater Creek, across Kneeland and down Lawrence Creek and Yager Creek, across US 101 to the small community of Stafford—half destroyed by a massive mudslide beginning in a PALCO road on the steep hillside above Stafford in Winter 1997. On a flight in September 1997, I witnessed the fresh clear cuts scarring steep slopes, silt-laden streams after the first rains of autumn, brush-covered hillsides killed by herbicides sprayed on them by PALCO to kill back hardwood species (including alders) and encourage industrial plantations of conifer trees.

During the 1990s, the battle over the ancient redwoods came to represent many conflicts in American culture at the end of the millennium—the battle between mega corporations and those who demand grassroot citizens control of democratic processes, the battle for empowerment of women who speak for the forest, the battle for empowerment of local citizens dwelling in watersheds, and social conflicts over anthropocentric versus ecocentric philosophy.

Alston Chase, in his book *In a Dark Woods*, asserts that the battle over the integrity of forests and watersheds in the Pacific Northwest, especially the battle over the old-growth redwoods in Humboldt County, is a battle over the philosophical bedrock of Western civilization. For Chase, this conflict is class warfare. "Promoted by government, activists, and scientists, biocentrism had become a philosophy of America's ruling classes—the media, government, universities, church hierarchy, and teaching professions" (Chase, 1995, p. 359).

For Chase, nature is resilient while civilizations are fragile. In Chase's narrative, the most misguided, dangerous theory in ecology is the theory of ecosystem stability. There is no "balance of nature," says Chase, and ecosystem management is just another term for preservation.

Because Chase's views are widely held among supporters of the "wise use" movement and other critics of the conservation movement, and because he, and others like him, see the battle over redwood forests as the final stand for Western civilization, it is important to clarify the issues from an ecocentric perspective.

Max Oelschlaeger in his scholarly book *The Idea of Wilderness* clearly distinguishes between resourcism, the dominant philosophy of modern industrial civilization, and ecocentrism.

Oelschlaeger concludes that resource conservationists believe that systems are no more than collections of parts. Homo sapiens are related externally to the ecomachine. The ecomachine can be engineered to produce desired outcomes and prevent undesired consequences. The market objectively determines the worth and value of all things, cultural and natural. The national and per capita income accounts are the ideal measure of societal well-being. Progress is determined according to the utilitarian formula of the greatest good for the greatest number.

Ecocentrists believe that natural systems are the basis of all organic existence, and therefore possess intrinsic value. Humankind is an element within rather than the reason to be of natural systems and is hence dependent upon intrinsic value. Ethical human actions (actions which promote the good life for humankind) necessarily promote all life on Earth (preserves such intrinsic values as diversity, stability, and beauty) (Oelschlaeger, 1991, p. 294).

In my interviews with forest activists, I find that in their personal ecosophy, they frequently take a mild holism approach rather than a hard holism approach. That is, they emphasize communal responsibilities and community stability rather than excessive egoism and individualism. Mild holism implies that the integrity of natural systems, allowing them to continue their evolutionary journey, is the context within which humans dwell. That which tends to preserve the integrity of endemic species in the bioregion and protect the integrity of natural processes within a range of historic variability is right and actions which tend otherwise are wrong.[5]

Many forest activists in the bioregion identify corporations as primary actors in the battle over the future of forests in the Pacific Northwest. Corporations, as a rule, continue to discount the future at ever more rapid rates. There is no evidence that any industrial logging corporation will ever grow an old-growth redwood forest. No industrial logging corporation has ever grown a thousand-year-old redwood tree or even a two-hundred-year-old redwood tree. Industrial logging corporations want to grow trees like farmers grow corn. They want to "harvest" tree flesh every forty years or even every twenty years. Instead of complex forests, industrial foresters want to convert the forested landscape into tree plantations. Even black bears, the great sacred animal of the forests, are only "industrial pests" to industrial foresters. Some black bears like to scratch the cambial layer of young redwood trees, perhaps for the sweetness of that layer during the spring season. Industrial foresters have asked the California Fish and Game Commission to declare black bears on their lands "industrial pests" so that they can allow hunters to kill black bears on their lands.

In contrast, ecocentrism represents a shift away from viewing humans and the human drama as central to the history of the cosmos to viewing humans, in the words of ecologist Aldo Leopold, as "plain citizens" of the cosmos. Our search, as humans, is to find a grounding for our lives in a deep sense of reverence and love for our bioregion. This search is a search for ecological wisdom or ecosophy.

Alan Drengson, a philosopher, provides an eloquent statement of the search for ecosophy as arising from our sense of place:

> Ecosophy is wisdom which arises from a place and resides in it, and it is also the wisdom to dwell harmoniously in the place. While ecosophy arises within a particular place, by showing us all of a place, it opens to unlimited space, and is global, universal and expansive. It deepens and evolves as a free creative presence in which multitudes of beings participate. It can be reached by way of the narrative and other trails. . . . Ecosophy is ecological wisdom and harmony involving comprehensive, compassionate appreciation for Nature with Its myriad communities of beings. As compassion, it is a wise sensibility that is harmonious in *ecos* with others. Ecosophy is ecological wisdom, pursued and realized through the practice of disciplines which bring harmony and understanding into our ecological relationships. (Drengson and Inoue, 1995)

Ecosophy flows from our intuitive interconnectedness with our bioregion and with understanding scientific principles of the evolution of ecosystems. In his analysis of the history of the concept of ecosystem management, "What is Ecosystem Management" (Journal of Conservation Biology, 8,1 27–38), Ed Grumbine concludes that there are five general goals of ecosystem management under the overall goal of sustaining ecological integrity. These are: (1) Maintain viable populations of all native species in situ; (2) Represent, within protected areas, all native ecosystem types across their natural range of variation; (3) Maintain evolutionary and ecological processes (i.e., disturbance regimes, hydrological processes, nutrient cycles, etc.); (4) Manage over periods of time long enough to maintain the evolutionary potential of species and ecosystems; (5) Accommodate human use and occupancy within these constraints.

There are values, deep in the woods. These values bear on the question "Do humans have the right to willfully cause the extinction of other species for their own egotistical reasons?" The answer, for an ecocentrist, is clearly NO. Our experience with the forests and salmon lead us to proclaim "Let the River Live."

Modern civilization has become an empire of resource extraction. Some logging corporations active in northwestern California in the late 1990s were also looking to Chile, New Zealand, and Siberia, among other places, as sources for more timber. From Siberia to Chile, forest destruction is occurring at an increasing rate in the pursuit of profits and economic growth in a globalizing economy.[6]

In September 1998, Governor Pete Wilson signed a bill authorizing the state of California to use public funds to match already approved federal funds to buy Headwaters Forest, Owl Creek, Grizzly Creek, and several other isolated groves of redwoods from Maxxam (PALCO). Only one conservationist attended the signing ceremony, the executive director of Save The Redwoods League. Ironically, the Save The Redwoods League was the one conservation group most isolated from the struggles over preservation of the Headwaters Forest complex from 1987 through 1998. As usual, the media and some politicians claimed that only a "radical fringe" of conservationists, including Sierra Club, EPIC, and some Earth First!ers, were displeased with the "deal" signed by Governor Wilson.

On March 1, 1999, the federal government signed an agreement with Maxxam (PALCO) to purchase Headwater Forest and other groves totaling 7,470 acres, including acres of uncut ancient forest. The agreement also provided fifty-year protocol of an additional 7,728 acres, including 1,446 acres of uncut ancient redwoods, and provided for potential acquisition by the state of California of more than 2,000 additional acres, including about 450 acres of uncut ancient forest.

While applauding the willingness of the federal and state government to spend a total of $495 million to acquire priceless old-growth redwood groves, many conservationists expressed anger that Charles Hurwitz again received a cash payoff for his blatant attack on the integrity of the redwood forests. They also expressed grave concern that the payments to Charles Hurwitz could establish a precedent whereby the government would pay compensation to landowners for protecting the habitat of threatened and endangered species.

Furthermore, the Habitat Conservation Plan (HCP), covering all of Maxxam's holdings in Humboldt County and lasting fifty years, lacked provisions for citizen intervention (through lawsuits) if Maxxam or a successor corporation failed to fulfill the goals of the HCP, lacked mechanisms for effective monitoring of the HCP over the life of the HCP, and basically allowed Maxxam to liquidate remaining, residual old-growth from its lands.

After the federal government signed the Headwaters Forest Agreement, a parade of public officials including Secretary of Interior Bruce Babbitt,

Senator Diane Feinstein, and the governor, Gray Davis, trooped to the Headwaters Forest for media photo-ops. Earth First!ers were not invited to hike to the Headwaters Forest with the governor.

In a press release, EPIC summarized the contributions of grassroots activists to the campaign for the Headwaters Forest.

> Without the past twelve years of grassroots action and citizen advocacy, this entire forest would have been turned into lumber long ago. Many of the politicians congratulating themselves for completing the Headwaters deal stood by silently while previous acquisition efforts—and thousands of ancient trees—fell crashing to the ground. Most of the people who truly deserve credit for achieving even the partial protection offered by this deal have not been named in the newspapers or in the politicians' press releases: the Earth First! activists who hiked, named ,and defended the groves; the volunteers and attorneys who worked countless hours to bring EPIC its landmark legal victories; the thousands of individuals and families who attended rallies, demonstrations, meetings, and hearings throughout California; the citizens who wrote their elected officials, sent letters to the editor, and supported the various efforts to save Headwaters; and the scientists who put their reputations on the line in order to speak the truth on behalf of the water, fish, and wildlife of Northern California. These are Headwaters Forest's true heroes, and we trust that they will not abandon the work that still remains to be done. Although it was taken from our hands at the very end, the effort to save Headwaters has always been a true grassroots uprising. In many ways, the last twelve years have created the groundwork for an increasingly informed and active public to succeed in changing the way we treat our forests, our water, and our remaining wild creatures. (EPIC Update March 22, 1999)

For bioregionalists, the greatest irony of the outcome of the battle over the Headwaters Forest complex was the increasing involvement of the federal government in forest and watershed management throughout the bioregion. Grassroots conservationists went to federal courts again and again from the late 1980s through the 1990s to force federal agencies to enforce provisions of the ESA, and although they were reluctant to act, federal agencies, ever since the end of the 1990s, have greater regulatory power over private timberlands in the bioregion and have been ordered by the courts to use those powers to protect the integrity of the habitat for northern spotted owls, marbled murrelets, coho salmon, and other threatened and endangered species in the bioregion.

Another ironic result of the outcome of twelve years of social conflict over the fate of the Headwaters Forest complex is that the general public will have very limited access to the groves of ancient redwoods acquired with public funds. Only the Grizzly Creek acquisition is adjacent to an existing state park and therefore available for recreation activities by the general public. The other groves acquired by the government under the "deal" are still surrounded by thousands of acres of private Maxxam (PALCO) lands. Federal and state regulatory agencies will have to monitor and supervise implementation of the HCP and Sustained-yield plan (SYP) which are part of the "deal."

Many questions remain. What role, if any, will grassroots conservationists play in monitoring compliance with the HCP? Will federal and state agencies be more aggressive than they have been during the past several decades in enforcing federal and state legislation designed to achieve conservation objectives on private lands? Will it be possible to restore viable wild populations of coho salmon and steelhead in the twenty-first century?

While all stakeholders in the Headwaters Forest controversies argue for a "science-based" approach to ecosystem management, bioregionalists argue that ecosystem management should mean more management of human activities and restraint on human desires to maximize short-term profits. Crafting a way of life in the bioregion will remain more of an art, and indeed a spiritual practice, than just a "science-based" strategy. Many grassroots conservationists continue to assert that without full public ownership of the 60,000 acres of the Headwaters Forest complex, the corporate objective of sustained profits will never be reconciled with the public interest in protecting the integrity, intrinsic values, and spiritual values of the forest and watershed system.

Certainly when Greg King and a few other young Earth First! activists discovered and named Headwaters Forest in 1987 they never expected that the battle over this forest, lasting over twelve years, would be the longest battle over redwood forest protection during the twentieth century. Nor did they realize the toll it would take on the lives of grassroots activists. Unlike the battle to create a Redwood National Park during the 1960s, no national conservation group or even statewide conservation group, such as Save the Redwoods League and the Planning and Conservation League, carried the burden of defending the Headwaters Forest complex.

Tragically, and ironically, the first casualty in the battles over the fate of the Headwaters Forest complex occurred a week after Governor Wilson signed the bill authorizing participation by the state of California in the purchase of Headwaters Forest. David "Gypsy" Chain, a young man from Texas, was participating with other Earth First!ers in an action near Grizzly Creek to protest logging in alleged marbled murrelet habitat. According to a video taken by another Earth First!er an hour before Chain's death, loggers chased Earth First!ers in the area of the active logging operation. According to reports by some Earth First!ers present in the forest at that time, a tree being felled hit another tree which fell on Chain, killing him instantly.

Perhaps, in the twenty-first century, some will look back and realize the sacrifices, dedication, commitment, and spiritual energy that grassroots activists devoted to this campaign. Perhaps the conservationists of the future will dedicate a monument to the named and unnamed activists in this struggle—beginning with Greg King and Larry Evans, continuing with Darryl Cherney and Judi Bari, and Peanut, Screech Owl, Spring, Gypsy, Josh Kaufman, and all the thousands of activists who were arrested while demonstrating, protesting, expressing their passion, love, and commitment to something greater than their own egos. They brought rich feelings to a world of fact. Their gift to the bioregion will be remembered by all who care.

From my many memories of demonstrations, rallies, protests, and celebrations in defense of Headwaters Forest complex, one personal memory symbolizes my feelings. On a winter day, early in 1996, driving home on rain-slick pavement from yet another rally for Headwaters, I stopped at Bull Creek, a tributary of the Eel River that was massively logged between 1947 and 1964 in the upper reaches of the watershed. The lower sections of the watershed are old-growth redwood forests within the boundaries of Humboldt Redwoods State Park. I walked to the Flatiron Tree, a massive old-growth redwood that fell during a winter storm in 1995. In the growing darkness of early winter, the moist forest was alive with energy. A fallen redwood giant is not dead. It is alive and is host to many species of fungi and moss. Standing next to the fallen giant I hugged it and began to cry—tears of joy. I am a tree hugger. I am proud to be a tree hugger. I stood with the fallen tree and with all tree huggers. I affirmed that trees have standing. I realized that I come from a long line of tree huggers, guardians of the forest, guardians dwelling in the dark woods. Huwawa, the guardian of the cedar forest in The Epic of Gilgamesh, Pan, calling us into the dark woods throughout the tragic, hubris filled history of Western civilization, the Druid guardians of the forests of Europe before they were destroyed by Christian invaders, Raven and the great bear guardians of the forests of the northern hemisphere, John Muir, David Brower and all the Earth First! tree sitters and protesters who have witnessed for the integrity of redwood forests.

In the dark woods I was a bear becoming human and a human becoming a bear. The sky bear, Ursus Major, is rising above this forest on this winter night. The earth bear is sleeping in her den, awaiting resurrection in the spring. In the dark woods I am home, bearing up to the suffering, bearing witness for the integrity of myself as part of the forest, bearing forth in the silence of the night with energy flowing through the forest, with the intelligence of the forest. Deep in the dark woods is the spirit of life. Rivers flow through my veins and the ocean swells within my body. Coming to my senses, deep in the dark woods, I knew that I was home.

During the first decade of the twenty-first century, the Bush administration attempted to roll back many of the conservation efforts of the Clinton administration. One of the largest private landowners, Green Diamond Corporation, with headquarters in the state of Washington, signed fifty-year HCPs with the federal government. One plan covering 416,000 acres in Humboldt County is based on the intent to keep sediment from entering state waters. This HBP covers fisheries and amphibians. The success of these HCPs depends on commitment of private landowners over several generations and willingness of federal agencies over several administrations to monitor the actions of private landowners.

Activities on public lands, especially those administered by the U.S. Forest Service, including Six Rivers, Mendocino, Klamath, and Shasta-Trinity National Forests, are also based on political decisions made in Washington, DC, While the U.S. Forest Service says it will make decisions based on the "collaboration" of local residents, the final decisions are made by U.S. Forest Service employees.

In summary, political power remains in the hands of the Federal government and large private corporations.

NOTES

1. For a detailed historical and sociological account of the timber wars during the late twentieth century on the northcoast of California, see Widick (2009). For an overall view of radical environmentalism in relation to natural resource extraction, see Green Rage by Christopher Manes (1996).
2. David Harris (1996) has written specifically on the events and personalities surrounding the corporate takeover of the PALCO by Charles Hurwitz of the Maxxam Corporation, and latter's corporate moves to clear-cut the ancient forests of Douglas fir and coastal Redwoods on private lands owned by PALCO.
3. Redwood Summer occurred in the summer of 1990—there was concerted action by forest activists, who attempted to block logging roads, tree sit, etc. For specific accounts and actions of this summer of protests, see Bari (1993), List (1993), and Pickett (1993)
4. One of the most notorious act of tree sitting was by Julia Hill (2001).
5. The work of Leopold (1949) discusses this theme in a more extensive context.
6. Global deforestation has been an issue throughout world history. A global assessment of the impact of deforestation on world development was made by the World Commission on Forests and Sustainable Development (1999) For in-depth studies see Chew (2001), Dauvergne (1997), Lensky (1992), and Mendes (1990).

Chapter 8

Fire on the Mountain

Fire is a primal force. Fire helped create the mosaic of forests in the Klamath-Siskiyou bioregion. Deep in old-growth coastal redwood forests, we find huge fire scars on ancient redwoods, yet the trees still live. Massive lightning-caused forest fires have raged through the Salmon River, Dillon Creek, Pilot Creek, and portions of the Trinity Alps Wilderness during the last thirty years.

Fire is also an essential aspect of human culture. We sing around the campfire. We gather wood to burn in our fireplaces. Warmed by burning tree flesh during long winter nights we gaze into the flames, flickering blue and orange, looking for the phoenix arising. Yet we also fear fires, fear that our homes will burn, fear that firefighters will lose their lives. Ashes to ashes and out of the ashes new life emerges.

In its tradition of managing nature, the U.S. Forest Service (USFS) developed "fire management plans" for each national forest. For nearly eighty years the USFS has been in the business of fire suppression—at any cost. Smokey the Bear was created to promote the message "only you can prevent forest fires." Yet during the 1990s, more and more fire ecologists argue that fire is a dynamic, creative positive force in the evolution of a landscape and that unrelenting fire suppression may place the forest at greater risk of a catastrophic fire sometime in the future.[1]

Going to the mountains, to the Klamath River settlement of Orleans, I listened to learned scholars and scientists proclaim their new vision of fire on the mountain. Some of the speakers at this fire ecology symposium, held in November 1997, emphasized that as a human community we are in a period of transition in our approach to forest fires. In this bioregion, and many other regions of the American West, managers, landowners, and grassroots conservationists are moving from the view that we must always suppress fire, especially at the urban/wild lands interface, to the view that we must understand the uses of fire as a tool of "ecosystem management?" How can science help us understand the role of fire in landscape evolution? What values do we have concerning forest fires? How do ecologists see fires as a regenerating force in vegetation and wildlife habitat? How can we balance our love of fire, pyrophilia, with our fear of fire, pyrophobia.

The Klamath-Siskiyou province as a whole is diverse in soil and rock types and in forest, wildlife, and plant communities. Various watersheds and elevations in the mountains of the Klamath-Siskiyou province also exhibit wide variations in fire disturbance regimes. It is difficult to make generalizations about the role of fire in the coastal and interior mountains of northwestern California and southwestern Oregon.

Lightning-caused fires during the late summer and early fall have moved across the landscape of these mountains for millennia.

When contemporary scientists view forest fires, they do so with an attempt to describe the characteristics of fires and predict the consequences of fires on the landscape.

A fire ecologist does not consider the commodity values of the forest in studying the effects of forest fires. Forest ecologists ask scientific questions such as how do plants and wildlife adapt to fire regimes? What is the fire regime in the Klamath-Siskiyou bioregion? How do we describe a forest fire as a disturbance event? A disturbance event is a discrete event in time that has a significant effect on species composition, forest structure and function, and habitat of various species of wildlife. The fire ecologist wants to quantify disturbance frequency and variability. That is, in the history of the landscape, where does a given forest fire fall in the range of historical variability? Is a given forest fire an extreme event or an average event of this type?

Considered as a disturbance type event, forest fires vary in their magnitude and extent. *Magnitude* refers to varying intensities/severities of fire. *Extent* refers to quantitative measures of the duration (time) of the fire and space (distance fire traveled, acres burned). Most news stories on forest fires cite the acres burned but not the magnitude of the burn. Small patches can be burned with low, medium, or high intensity and larger patches of the forest can be burned with low intensity. In some cases due to changing humidity levels, wind speed, wind direction, time of day, topography, vegetation, and temperature, among other factors, a fire may skip across the landscape. The extent of the fire can range from low intensity burn, where only under story vegetation is killed by the fire, to very high intensity burn where most or all of the standing green trees are killed from the burning. The season during which the forest fire occurs will determine the extent of disturbance and magnitude on endemic or migratory wildlife that use the landscape for nesting or feeding.

Fire is seen by fire ecologists as one of the forces that allows the evolution of a landscape to continue. In the "new forestry" approach, there are no old-growth forests. There are "shifting gap phases successional forests." Blowdowns create gaps in the forest structure. Low intensity forest fires create gaps in the forest structure. The gaps are filled first with brush and then with young conifer trees. Pyrodiversity enhances biodiversity.

Forest ecologists over the past generation have discovered the role of dead wood, snags, and downed trees as habitat in the forest. Forest fires help create large snags and logs, for example. Forest ecologists have also developed indices of the historic range of variability of forest fires, floods, wind storms, and other naturally occurring disturbance events in the forest landscape. Landscapes where human interference has been minimal could provide benchmark data that can be compared with data from landscapes that have experienced extensive modification during the past century by grazing, logging, and fire suppression. It is a societal decision whether nature should take its own course in relatively pristine landscapes, such as designated wilderness areas, and in extremely human-modified landscapes, such as the tree plantations that have replaced many of the complex forest systems in this bioregion.

However, what should be our approach to landscapes that have been extensively modified by humans, especially those forest/mountain landscapes that have been extensively modified by forest management practices during this century?

Reed Noss, forest ecologist and consultant to the Wildlands Project, says that "letting nature take its course is unethical in areas that humans have impacted. We have a responsibility to help the landscape in its healing process to the extent that we are able with our current wisdom." Within the ethical dimension of decision-making, the forest ecologist can use credible science to help solve real problems. Credible science should stand up to peer review and stand up in court. He argued that scientists are not activists. Scientists come up with options or ways to achieve goals determined in the political process of a society. The selection of which option to apply to public policy is a political, not a scientific, decision.

In the contemporary rhetoric of forest management, industrial foresters, public lands foresters, and forest activists use the language of "forest health" and "protecting the integrity of the forest." "We must live in harmony with nature." "We want to live in balance with nature." "We want sustainable forests and we can have both continued economic growth and sustainable forests." Feminists talk of the "partnership" between humans and forests much as they talk about "partnership" between men and women, a relationship in which each partner has equal power.[2]

Some progressive foresters in the social forestry movement talk of a "stewardship model" for our relationship with forests. The steward in Christian theology is appointed to represent an absentee landlord (God) and protect the property of the absentee landlord.

In the secular version of the "stewardship model," the forest manager protects the integrity of the fully functioning forest system for the continued well-being of present and future generations of humans. When the stewardship model is based on secular economics, not on moral authority, decisions concerning management of the forest are usually made based on short-term economic considerations. The integrity of the forest/watershed system can always be compromised in order to log timber to pay for social welfare costs, provide jobs for timber workers, and provide "necessary products" for the human population.

However, place-based spirituality involves us in a delicate interpenetration of mind and nature. Nature is not a factory producing sustenance for humans. Nature is not a background upon which the human drama is played. Nature is not managed for predetermined goals or desired future states. Nature is. And humans interpenetrate with the "isness" of Nature through our constant attention to change. The continuous passing away and continuous arising of being commands our attention.

Indians living along the Klamath and Trinity Rivers traditionally used fire, but the extent to which they deliberately burned forests in the Klamath-Siskiyou province remains an open question. The Karuk nation lives in the upper Klamath River watershed. In Karuk religion, humans were given the responsibility by the Great Creator to keep the world in balance. Rocks, trees, and wildlife are all considered "our relations," our cousins. As we lose these other beings, we lose who we are. We reciprocate with all our relations. Recently we haven't looked after them so they haven't looked after us.

In their article "Environmental Management: American Indian Knowledge and the Problem of Sustainability" (*Forest, Trees and People Newsletter* No. 34), Leaf Hillman and John F. Salter conclude that

> the Karuk have at a central place in their attitudes toward nature and sacred ceremonies, the profound sense that the role of this tribe is to fix the world each year, making right the wrongs of the past seasons. Traditionally this has referred to reaffirming behavior and perspectives which promote concerned and sustainable relations with nature. Having evolved over an immense period of time, Karuk land management finds multiple expression-environmental knowledge, technical and ritual practices, underlying attitudes toward nature and a conception of the role of humans in a natural system which is one of responsibility to all other living creatures. One aspect of this sense of the sacred is that one would never lie about or misuse a sacred relationship or responsibility. Similarly, the sacred associations and affinities held by the Karuk for the ancestral landscapes translate into that which would never be slighted or abandoned.

How can place-based spirituality concerning our attention to forest fires be reconciled with "strategies for public lands fire management at the landscape level?" Are we justified in our fear of catastrophic forest fires either caused by natural forces such as lightning or human caused as with arson fires? Must we log the forest in order to save it from catastrophic fire?

Fire on the mountain, fire in the heart. On clear days during November, "permissible burn days," loggers are burning slash fires in the hills above Arcata. After logging, bulldozers are used to push slash or saplings and brush killed during the logging operation into piles that are then burned.

In the brush fields of the Klamath-Siskiyou bioregion where great forests stood only decades ago, fire will come, and move on. And visitors, many centuries in the future coming into this landscape, will see a landscape affected by the decisions we make during this decade concerning our relationship with forest fire.

NOTES

1. See World Commission on Forests and Sustainable Development (1999) on the use of controlled burns as a forest management tool.

2. For explication of ecofeminism, see Cheney (1987), Mies and Shiva (1993), Merchant (1992), King (1992), and Warren (1992).

Chapter 9

The Republic of Ecotopia

During the twentieth century, the conservation movement provided the only opposition to the Party of Progress and Economic Growth.

There is no doubt that without the strenuous work of the conservation movement over the past four decades in the Klamath-Siskiyou region there would be no old-growth forests remaining in this bioregion and the integrity of the natural systems would be more threatened than they are now. There would be no regulatory structure that has the potential to encourage loggers, miners, grazers, and fishermen to be responsible for their actions.

During the first decade of the twenty-first century, deeper questions are being asked. What does it mean to dwell in a region rather than only extract resources from a region? What is life value rather than the dollar value of logs in the marketplace?

Bioregionalism, as a crafted way of dwelling, within the limits of a region, is more widely accepted in the general population.

The real work for the next generation of conservation activists will be to build on the achievements of conservationists during the last four decades of the twentieth century. Writer Ernest Callenbach coined the term "ecotopia." In his novels *Ecotopia* (1975) and *Ecotopia Emerging* (1982), he placed his "back-to-the-land" philosophy in northwestern California.

Emerging trends and ecological changes in this region during the past four decades illustrate the possibility of a bioregional culture in the Klamath-Siskiyou region.

Among changes already underway is the removal of dams. Hundreds of licenses to operate dams are up for renewal during the next several decades. "Let the rivers live" has already moved from slogan to public policy. The continuing campaign to remove two dams on the upper Klamath River, in Oregon, to allow salmon to return to their ancestral spawning grounds illustrates a change in public attitudes toward dwelling with the flow of rivers rather than controlling rivers.

The removal of roads on public lands in the region is widely practiced. Federal agencies managing public lands in the region including the National Park Service, the US Forest Service, and the BLM, have embarked on large-scale removal of thousands of miles of logging roads.

During the era of global warming, forests are seen as vast sequesters of carbon and therefore protecting forests of the Klamath-Siskiyou region from massive logging is one aspect of the changing policy toward dwelling in this region.

Under provisions of legislation signed by the governor of California in 2006, the State Air Resources Board is required to design and implement strategies to reduce CO_2 emissions in California. Forests play an important role in emerging strategies.

Conservation groups are developing strategies and programs to protect native biodiversity. Land trusts have been established in many parts of this region to maintain conservation easements on private lands in order to maintain ranching and farming.

The state of California, through the Resources Agency, has adopted a bioregional approach to policy. The state of California encourages a bottoms-up approach to planning rather than a top-down approach. The state of California encourages various cooperative approaches where all stakeholders—including federal and state agencies, Indian tribes, county governments, conservation organizations, ranchers, and logging corporations—have been asked to develop long-term integrated management objectives.

Watershed associations have been formed on many rivers and streams. A new social order is emerging. We cannot predict what shape these watershed associations will take over the next several decades, but most likely the emerging social order will be very different from the existing command and control model of regulatory action.

Some established conservation groups have adopted a systems approach to planning their own initiatives for the twenty-first century. For example, the Save The Redwoods League has commissioned both a scientific report on the state of the redwood biome and a survey of major stakeholders, including conservationists and landowners, in the redwood region. The League will use these documents in developing an integrated strategy for the work of the League into the twenty-first century.

Political leadership is required to speak for political autonomy and localization.

At the beginning of the twenty-first century there are many environmental organizations active in the Klamath-Siskiyou bioregion. A partial list includes Friends of the Eel River, Legacy: The Landscape

Connection, Citizens for Better Forestry, Northcoast Environmental Center, EPIC, The Trees Foundation, Redwood Region Audubon, Conservation Unlimited, Sierra Club: North Group, Institute for Sustainable Forestry, Kalmiopsis Audubon, Friends of the Dunes, Jacoby Creek Land Trust, Green Party, Students for Ethical Treatment of Animals, Taxpayers for Headwaters, Friends of Del Norte, Karuk Fisheries, Mattole Restoration Council, Salmonid Restoration Federation, American Indian Alliance, Yurok Fisheries, South Fork Mt. Defense Committee, McKinleyville Land Trust, Arcata Community Recycling Center, Alliance for Paving Moratorium, Salmon Forever, Surfriders Foundation, Smith River Alliance, Center for Natural Lands Management, Cal-Trout, Seventh Generation Foundation, Mill Creek Conservancy, Sanctuary Forest, CCAT, and Siskiyou Mountains Resources Council.

It is time to announce a vision for the twenty-first century.

Let the rivers live. With that as their slogan, bioregionalists can begin to reinhabit, begin to dwell deeply in the way tribal peoples always dwelled, in the spirit of this bioregion. Let the rivers live. Free-flowing rivers, rivers that are the life, veins, and arteries of this bioregion, teach us humility and sustain us.

References

Abberley, Doug. 1999 "Interpreting Bioregionalism." in McGinnis, Michael (ed.), *Bioregionalism*, pp. 13–42. London: Routledge.

Arnold, Mary Ellicott and Mabel Reed. 1957. In *The Land of the Grasshopper Song*. Lincoln, NE: University of Nebraska Press.

Avery, Stephen. 2004. "The Misbegotten Child of Deep Ecology." *Environmental Values* 13(1): 31–50.

Bahaguna, Sunderlal. 1996. "Himalaya – From Disaster to Survival." In Jayant Lele and Wisdom Tettey (eds.), *Asia—Who Pays for Growth*, pp. 181–190. Aldershot: Dartmouth.

Bari, Judi. 1994. *Timber Wars*. Monroe, ME: Common Courage Press.

Becking, Rudolf W. 1982. *Pocket Flora of the Redwood Forest*. Covelo, CA: Island Press.

Bell, Maureen. 1991. *Karuk: The Upriver People*. Happy Camp, CA: Naturegraph Publisher.

Berg, Peter. 1978. *Reinhabiting a Separate Country*. San Francisco, CA: Planet Drum Foundation.

———. 2009. *Envisioning Sustainability*. San Francisco, CA: Subculture Books.

Berg, Peter and Dasmann, Ray. 1978. "Reinhabiting California." in Berg, Peter (ed.), *Reinhabiting a Separate Country*, pp. 217–220. San Francisco, CA: Planet Drum Foundation.

Berry, Thomas. 1988. *The Dream of the Earth*. San Francisco, CA: Sierra Club Books.

Callenbach, Ernest. 1975. *Ecotopia*. Berkeley, CA: Bayan Tree Books.

Carr, Mike. 2005. *Bioregionalism and Civil Society: Democratic Challenges to Corporate Globalism*. Vancouver, BC: University of British Columbia Press.

Chase, Alston. 1995. *In a Dark Wood: The Fight over Forests and the Rising Tyranny of Ecology*. New York: Houghton Mifflin.

Cheney, Jim. 1987. "Ecofeminism and Deep Ecology" *Environmental Ethics* 9: 115–145.

Chew, Sing C. 1997. "For Nature: Deep Greening World-Systems Analysis for the 21st Century." *Journal of World-Systems Research* 3(3): 381–402.

———. *World Ecological Degradation: Accumulation, Urbanization and Deforestation 3000 BC-AD2000*. Walnut Creek, CA: Alta Mira Press.

———. 2008. *Ecological Futures: What History Can Teach Us*. Walnut Creek, CA: AltaMira Press.

Cohodes, Marvin. 1997. *Basket Weavers for the California Curio Trade: Elizabeth and Louise Hickox*. Tucson, AZ: University of Arizona Press.

Coile, Zachary. 1999. "You Say You Want a Revolution...2 1/2 years after Green Party won in Arcata, Changes Are Less Than Radical." *San Francisco Examiner*, April 18.

Cooperrider, Allen, Steven Day and Curtice Jacoby. "The Bioreserve Strategy for Conserving Biodiversity." In Richard K. Baydack, Henry Campa III, and Jonathan B. Haufle (eds.), *Practical Approaches to the Conservation of Biodiversity*, pp. 35–53. Covelo, CA: Island Press.

Curtius, Mary. 1998. "Greens Wilt at City Hall." *San Jose Mercury News*, October 12, 3B.

Dasmann, Raymond. 1965. *The Destruction of California*. New York: Macmillan.

Dauvergne, Paul. 1997. *Shadows in the Forest*. Boston, MA: MIT Press.

Davidson, Stewart. 2007. "The Troubled Marriage of Deep Ecology and Bioregionalism." *Environmental Values* 16(3): 313–332.

Delong, John Raymond. 1968. "The Ravaged Remnant: A History of the Logging and Preservation of the California Coast Redwood." M.A. Thesis, San Francisco State College.

Dengler, Lorinda Ann. 1992. "On Shaky Ground: Living with Earthquakes on the NorthCoast." Humboldt State University Special Collections and Archives. Arcata, CA.

———. 1995. Living on Shaky Ground. How to Survive Earthquakes and Tsunamis on the Northcoast." Humboldt State University Special Collections and Archives. Arcata, CA.

Devall, Bill. 1988. *Simple in Means, Rich in Ends: Practicing Deep Ecology*. Salt Lake City, UT: Peregrine Smith.

———. *Living Richly in an Age of Limits*. 1993. Salt Lake City, UT: Peregine Smith.

———. *Clear Cut: The Tragedy of Industrial Forestry*. San Francisco, CA: Sierra Club Books.

———. 2002. "The Deep, Long-Range Ecology Movement: 1960-2000—A Review." *Ethics and the Environment* 6(1): 18–41.

———. 2005. "Living in Mixed Communities of Humans, Condors, Mountain Lions, Bears, and Wildfires." *The Trumpeter: Journal of Ecosophy* 22(1): 10–28.

———. 2006. "The End of American Environmentalism?" *Nature and Culture* 1(2): 157–180.

Devall, Bill, Alan Drengon, and Harold Glasser, eds. 2005. *The Selected Works of Arne Naess*. The Netherlands: Springer.

Devall, Bill and George Sessions. 1985. *Deep Ecology: Living as if Nature Mattered*. Salt Lake City, UT: Peregrine Smith.

Dodge, Jim. 1981. "Living by Life: Some Bioregional Theory and Practice." *CoEvolution Quarterly* 32: 6–12.

Dolezal, Joshua A. 2008. "Literary Activism, Social Justice, and the Future of Bioregionalism. (Critical Essay)." *Ethics and the Environment* 13(1): 1–22.

Drengson, Alan and Yuichi Inoue. 1995. *The Deep Ecology Movement: An Introductory Anthology*. Berkeley, CA: North Atlantic Press.

Drengson, Alan and Duncan Taylor (eds.) 1997. *Ecoforestry: The Art and Science of Sustainable Forest Use*. Gabriola Island, BC: New Society Publishers.

Durbin, Kathie. 1996. *Tree Huggers*. San Francisco CA: Mountaineers Books.

Foreman, Dave. 1991. *Confessions of an Eco-Warrior*. New York: Harmony Books.

———. 1992. *The Big Outside*. New York: Harmony Books.

———. 2004. *Rewilding North America*. Washington DC: Island Press.

———. 2011. *Man Swarm and the Killing of Wild Life*. New York: Ravens Edge Press.

Fox, Warwick. 1990. *Toward a Transpersonal Ecology*. Boston, MA: Shambala Press.

Friedrichs, Jorg. 2013. *The Future Is Not What It Used To Be*. Boston, MA: MIT Press.

Garforth, Lisa. 2005 "Green Utopias: Beyond Apocalypse, Progress, and Pastoral." *Utopian Studies* 16(3): 393–427.

Geniella, Mike. 1997. "Power Struggle in Humboldt." *Santa Rosa Press Democrat*, December 22.

Grumbine, R. Edward. 1992. *Ghost Bears*. Covelo, CA: Island Press.

———. 1994. "What Is Ecosystem Management?" *Conservation Biology* 8(1): 27–38.

Harris, David. 1995. *The Last Stand: The War Between Wall Street and Main Street Over California's Ancient Redwoods*. New York: Times Books.

Harris, Stanley. 1996. *Northwestern California Birds*, 2nd edition. Arcata, CA: Humboldt State University Press.

Hart, John. 1975. *Hiking in Bigfoot Country: The Wildlands of Northern California and Southern Oregon*. San Francisco, CA: Sierra Club Books.

Heinberg, Richard. 2005. *The Party's Over*. Gabriola Island, BC: New Society Publishers.

Heizer, Robert and John E. Mills. 1952. *The Four Ages of Tsurai: A Documentary History of the Indian Village on Trinidad Bay*. Berkeley, CA: University of California Press.

Hill, Julia. 2001. *The Legacy of Luna*. New York: Harper.

Holland, Gal. 1996. "Town decides that green is its color: Environmentalists run Arcata, California." *USA Today*, November 18.

Hoopes, Chad L. 1971. *Lure of the Humboldt Bay Region*, revised edition. Dubuque, IA: Kendall/Hunt Publishing Company.

House, Freeman. 1999. *Totem Salmon: Life Lessons from Another Species*. Boston, MA: Beacon Press.

Hyde, Philip and Francois Leydet. 1963. *The Last Redwoods: Photographs and Story of a Vanishing Scenic Resource*. San Francisco, CA: Sierra Club Books.

International Panel on Climate Change. 2007. *Climate Change 2007*. Geneva: IPCC.

Isachson, Noah. 1999. "In Arcata, Medical Pot is no Big Deal." *San Francisco Examiner*, April 11.

King, Ynestra. 1992. "Toward and Ecological Feminism and a Feminist Ecology." in Peter List (ed.), *Radical Environmentalism*, pp. 70–80. Belmont, CA: Wadsworth.

Kneale Gould, Rebecca. 2005. *At Home in Nature*. Berkeley, CA: University of California Press.

Kroeber, Theodora. 1959. *The Inland Whale*. Berkeley CA: University of California Press.

LaChapelle, Dolores. 1988. *Sacred Land Sacred Sex: Rapture of the Deep*. Durango, CO: Kivaki Press.

Lanphere, Hortense M. 1990. *Bedlam on the Slew*. Lawrence, KS: Allen Press.

Lansky, Mitch. 1992. *Beyond the Beauty Strip: Saving What's Left of Our Forests*. Gardiner, ME: Tilbury Home.

Leopold, Aldo. 1949. *A Sand County Almanac*. New York: Oxford University Press.

Lewis, Oscar. 1966. *The Quest for Qual-a-wa-loo, Humboldt Bay: A Collection of Diaries and Historical Notes Pertaining to the Early Discoveries of the Area now Known as Humboldt County, California*. Oakland, CA: Holmes Book Co.

Leydet, Francois. 1969. *The Last Redwoods, and the Parkland of Redwood Creek*. San Francisco, CA: Sierra Club.

List, Peter (ed.). 1993. *Radical Environmentalism*. Belmont, CA: Wadsworth.

Lorentzen, Bob. 1996. *The Hiker's Hip Pocket Guide to the Humboldt Coast*, 2nd edition. Mendocino, CA: Bored Feet Publications.

Lowan-Trudeau, Gregory. 2014. "Considering Ecological Métissage: To Blend or Not to Blend?" *Journal of Experiential Education* 37(4): 351–366.

Macy, Joanna. 1991. *World as Lover, World as Self*. Berkeley, CA: Parallax.

Manes, Christopher. 1996. *Green Rage*. Boston, MA: Little Brown and Company.

Maser, Chris. 1992. *Forest Primeval: The Natural History of an Ancient Forest*. San Francisco, CA: Sierra Club Books.

———. 1992. *Global Imperative: Harmonizing Culture and Nature*. Walpole, NH: Still Point.

———. 1994. *Sustainable Forestry: Philosophy, Science and Economics*. Boca Raton, FL: St. Lucie Press.

Meadows, Donella. 1972. *The Limits to Growth*. New York: Signet.

———. 1992. *Beyond the Limits*. Post Mills, VT: Chelsea Green.

———. 2004. *Limits to Growth*. White River Junction, VT: Chelsea Green.

Mendes, Chico. 1990. *Fight for the Forests*. New York: Monthly Review Press.

Merchant, Carolyn. 1992. "Ecofeminism and Feminist Theory." In Peter List (ed.) *Radical Environmentalism*, pp. 49–54. Belmont: Wadsworth.

Mesarovic, Mihalo and Eduard Pestel. 1974. *Mankind at the Turning Point: The Second Report to the Club of Rome*. New York: Dutton.

Mies, Maria and Shiva Vandana. 1993. *Ecofeminism*. London: Zed Books.

Naess, Arne. 1973. "The Shallow and the Deep, Long-Range Ecology Movements: A Summary." *Inquiry* 16(1): 95–100.

———. 1992. *Crossing the Stones: A Portrait of Arne Naess*. Oley, PA: Bullfrog Films, Inc.

National Park Service. 1964 *The Redwoods: A National Opportunity for Conservation and Alternatives for Action*. Washington, D.C.: U.S. National Park Service.

———. 1998. *Redwood National Park Draft Management Plan and Environmental Impact Statement*. Washington, D.C.: U.S. National Park Service.

Nelson, Byron, Jr. 1978. *Our Home Forever: A Hupa Tribal History*. Hupa, CA: Hupa Tribe.

Norse, Eliott. 1990. *Ancient Forests of the Pacific Northwest*. New York: Island Press.

Norton, Jack. 1979. *Genocide in Northwestern California*. San Francisco, CA: Indian Historian Press.

Oelschlaeger, Max. 1991. *The Idea of Wilderness: From Prehistory to the Age of Ecology*. New Haven, CT: Yale University Press.

Pickart, A. J. and J. O. Sawyer. 1998. *Ecology and Restoration of Northern California Coastal Dunes*. Sacramento, CA: CNPS Press.

Pickett, Karen. 1993. "Redwood Summer Retrospective." In Peter List (ed.), *Radical Environmentalism*, pp. 207–212. Belmont, CA: Wadsworth.

Poole, William. 1998."Inside Ecotopia." *Sierra*, January/February, 30–33.

Raphael, Ray. 1974. *An Everyday History of Somewhere*. New York: Knopf.

———. 1985. *Cash Crop*. Mendocino, CA: Ridge Times Press.

———. 1981. *Tree Talk: The People and Politics of Timber*. Covelo, CA: Island Press.

Rohde, Jerry and Gisela. 1992. *Humboldt Redwoods State Park: The Complete Guide*. Eureka, CA: Miles and Miles.

Rodhe, Jerry and Gisela. 1994. *Guidebook to Redwood National and State Parks*. McKinleyville, CA: Mountain Home Press.

Rogers, Patrick. 1997. "Green Acres: Political guerrillas in Northern California must have taken charge of 'eco-groovy' Arcata." *People*, March 17, 93.

Roscoe, James. 1977. *The Mattole Valley: Economic Survival in a Rural Community*. Arcata, CA: James Roscoe.

Roszak, Theodore. (ed.) 1990. *Ecopsychology: Restoring the Earth healing the Mind*. San Francisco, CA: Sierra Club Books.

Sale, Kirkpatrick. 1991. *Dwellers of the Land*. Philadelphia, PA: New Society Publishers.

Schrepfer, Susan. 1983. *The Fight to Save the Redwoods: A History of Environmental Reform 1917-1978*. Madison, WI: University of Wisconsin Press.

Seed, John and Joanna Macy. 1988. *Thinking Like a Mountain: Towards a Council of All Beings*. Philadelphia, PA: New Society Publishers.

Sessions, George, ed. 1995. *Deep Ecology for the 21st Century: Readings on the Philosophy and Practice of the New Environmentalism*. Boston, MA: Shambala.

Snyder, Gary. 1974. *The Practice of the Wild*. San Francisco, CA: Northpoint Press.

———. 1980. *Turtle Island*. New York: New Directions Books.

Stillman, Peter G. 2010. "Educating for Ecological Sustainability in Political Science." *Ometeca*, 252.

Thayer, Robert L. 2003. *Life-Place*. Berkeley, CA: University of California Press

Trees Foundation. 1997. *Headwaters Stewardship Plan*. Garberville, CA: Trees Foundation.

Wallace, David Rains. 1983 *The Klamath Knot: Explorations of Myth and Evolution.* San Francisco, CA: Sierra Club Books.

Walmsley, Jesse. 1998. "Pretrial Publicity: Pepper Spray, Guilt Attribution, and the Change of Venue Motion." Senior Thesis, Department of Sociology, Humboldt State University, Arcata, CA.

Warren, Karen J. 1992. "The Power and the Promise of Ecological Feminism." In Peter List (ed.), *Radical Environmentalism.* Belmont, CA: Wadsworth.

Waterman, Thomas Talbot. 1965. *Yurok Geography.* New York: Kraus Reprint Corporation.

Wildes, Fred T. 1995. "Recent Themes in Conservation Philosophy and Policy in the United States." *Environmental Conservation* 22(2): 143–150.

Widick, Richard. 2009. *Trouble in the Forest: California's Redwood Timber Wars.* Minnesota, MN: University of Minnesota Press.

Witoszek, Nina, and Martin Lee Mueller. 2017. "Deep Ecology." *Worldviews* 21(3): 209–217.

World Commission on Forests and Sustainable Development. 1999. *Our Forests Our Future.* Cambridge: Cambridge University Press.

Index

About the Author and Editor

ABOUT THE AUTHOR

Bill Devall was in the Department of Sociology, Humboldt State University, from 1968 to 1995 and Professor Emeritus from 1995 to 2009. He was also Visiting Professor at the University of Alberta, Simon Fraser University, Australian National University, and the Naropa Institute in Boulder, Colorado.

He is the author of numerous articles and books: *Deep Ecology* (with George Sessions, 1985), *Simple in Means, Rich in Ends* (1988), *Living Richly in an Age of Limits* (1992), *Clear Cut: The Tragedy of Industrial Logging* (1993), and *Ecology of Wisdom* (with Alan Drengson, 2008).

ABOUT THE EDITOR

Sing C. Chew is the founding editor of the interdisciplinary journal, *Nature+Culture*. He is in the Department of Urban and Environmental Sociology at Helmholtz Centre for Environmental Research-UFZ, Leipzig, Germany, and Professor Emeritus at Humboldt State University. His books include *World Ecological Degradation* (vol. 1), *The Recurring Dark Ages* (vol. 2), and *Ecological Futures* (vol. 3). His most recent book is *The Southeast Asia Connection: Trade and Polities in the Eurasian World Economy 500 BC–AD 500.*

ABOUT THE BOOK

This book manuscript was completed by Bill Devall but was never published. An earlier version of the manuscript had the title *Humboldt County: Bioregion on the Edge*. Following comments from his friends and colleagues, Bill Devall pared the manuscript down and retitled it *Place and Life: A Bioregional Journey*. His health issues toward the end of his life prevented the publication of the manuscript.

Following the publisher's suggestion, the book manuscript has been retitled *Living Deep Ecology: A Bioregional Journey*. The notes in this book have been written by the editor so that the body of the manuscript is not interrupted by comments and bibliographic materials that were not in the completed manuscript. These bibliographic additions have been added for the reader's information.